"Scientist of the Islamic Era" is a book series encompassing eight volumes, the present book is volume 4 titled "Religious Scientists of the Islamic Era". It covers 84 religious Scientists encompassing people of tafsir and ta'wil, ilm-al-kalam, hadith, fiqh, justice, ijtihad, and ruhaniyyaat. The period of coverage is part 1 of the Islamic Era, from AD 610 to 1400. They commanded exceptional breadth in their learning and deepest insights in their specializations. They formulated the foundations of the fields of knowledge in the Religious Sciences and defined its frontiers. The field of religious science is unique to Muslims and Islam because Islam is a religion free from dogmas and therefore amenable to scientific treatment; and consequently Muslims were free to develop the required scientific world-view to formulate the related sciences.

Each scientist is briefly described. First, the name of the scientist is disambiguated, and an attempt is made to correct the misrepresentations common in the European translations. Salient scientific contributions of each scientist are briefly highlighted, a difficult task because of the fact that many of these scientists were polymaths. For each scientist we have provided a biographical summary to help picture their love and craving for knowledge, and the motivations and opportunities for them to do their research.

It is our objective that this fourth volume in the series will inform the Muslims about the wealth of their scientific heritage, and the next generations will feel inspired to surpass the excellence of their ancestors to enrich their heritage further, and be, like their ancestors, the flag bearers of world civilization in the religious sciences.

Muslims are now excelling in research with superb agility, and this series on Scientists of the Islamic Era will further stimulate this ongoing Renaissance in the Muslim world.

Religious Scientists of the Islamic Era

Volume 4 of the 8-volume series on

Scientists of the Islamic Era

Religious Scientists of the Islamic Era

Volume 4 of the 8-volume series on

Scientists of the Islamic Era

Abdur Rahim Choudhary, Ph.D.

Muslim Voice

MV Publishers

Published by MV Publishers, a subsidiary of Muslim Voice,
12719 Hillmeade Station Dr, Bowie, MD 20720, USA.
MVPublishers@muslimvoice.org

ISBN 978-1-956601-22-0

First edition 2025

United States of America

Choudhary, Abdur Rahim, 1944–
8-Volume Series on Scientists of the Islamic Era,
Volume 4, Religious Scientists of the Islamic Era

Muslim Voice

ISBN 978-1-956601-22-0

To the Muslim Ummah

Content

Preface to the 8-volume series on Scientists of the Islamic Era (Updated)

For a period of more than a millennium, Muslim Scientists have done foundational research in all scientific disciplines, and also greatly expanded the frontiers of science. However, our people often do not have a clear idea about our scientific heritage. We decided to write a series of books on "*Scientists of the Islamic Era*" that would be readily available to our generation and the coming generations, and provide motivation for excellence in the world civilizational dialogue, as well as to know our religious inspiration for scientific research and progression.

The young generations, especially those in Europe and Americas, have now opened their hearts and minds with a renewed desire for the truth about Islam and Muslims, being less influenced by historical biases and religious prejudices. The eight books in the series on *Scientists of the Islamic Era* seek to serve their youthful thirst for the truth.

Another reason for this series on "*Scientists of the Islamic Era*" is to produce a consciousness among the present-day academicians and scientists about the foundational contributions that the Muslim scientists made to all scientific disciplines, as well as how they expanded the frontiers of these disciplines. This fact is evidenced in the books in this series. However, this fact is not widely known because the present-day literature does not reference these original sources. The chain of scholarly references ends in European Renaissance, with occasional references to Greek scientists, but bypassing the millennium worth of research by Muslim scientists, who established the foundational principles and greatly expanded the frontiers of science.

In addition, the work seeks to fill a void, as no such series of books currently exists.

Islamic Era constitutes the period from 610 AD, when the Prophet received his first revelation, to 1922 AD, when the Ottoman Caliphate ended and the Turkish Republic began. We have divided the period in two parts: part 1 from 610 to 1400, and part 2 from 1400 to1922. The era is divided at an epoch when much of the works by the Muslim Scientists had already been translated into European languages, had become widely available, and had begun to produce Renaissance in Europe.

Each of the two parts of the Islamic Era is covered by the following four volumes, eight volumes in all.

Volume 1 is for Natural Sciences that include mathematics, astronomy, cryptoanalysis, chemistry, cartography, physics, and engineering based on these disciplines such as mechanics, automation, and robotics.

Volume 2 is for the Medical Sciences that include physicians, nurses, surgeons, herbalists, medical researchers, and medical writers.

Volume 3 is for the Social Sciences that include philosophers, historians, physical geographers, translators as well as the conventional sociology, political science, management sciences, economics, business, trade, anthropology, and linguists including letters.

Volume 4 is for the Religious Sciences that include Factologists, Mohadditheen, Qiyasists, Istihsanists, Qadhis, Mufassireen, Awwal-agists, Fuqaha, Mutakallimeen, Mujtahids, people of Ruhaniyaat, and Sufis. These terms will presently be explained.

Categorizing fields of scientific enquiry is currently done in the limited context of the European experience. For example, there are no words for the Ruh and Ruhaniyaat; people like Sufis; Muhadditheen who compile events of historic significance and fact check them for validity and accuracy; and the Faqih who exercise judgement using scriptural reasoning. In an attempt to classify the fields of research worldwide, it is necessary to use a classification scheme that is global and encompassing, rather than restrictive and limiting.

It therefore becomes necessary to introduce terminology in English language for, at least, the large and overarching fields of study that correspond to specialized Religious Studies in Islam. Corresponding terms for the Christian and Jewish religions exist in great detail incorporating their taxonomies to small granularities. For example, there are many terms in English language introduced to classify Catholic hierarchy within the Church. While the categories of people who study and practice Islam is vast, no terminology in English language has unto now been ascribed to describe them. Reasons for the circumstance are myriads but it is now time to permit a more expansive study of sciences of the Islamic Era. This requires to remove the gaps, and enlarge the research, make it global and free from omissions and restrictions. We have, therefore, introduced a minimal set of new terms for scientific studies of these topics. Below, these new terms are defined for use in the English language because no appropriate terms currently exist to represent the corresponding meanings.

Factologists

The term represents the scientists who investigate criteria used for the verification and validation of orally transmitted traditions or written historical statements. These were developed and practiced by the Muhadditheen. The science, among other things, uses what is known as 'knowledge about the people (Ilm-ur-Rijal)' encompassing all people who narrated or wrote down any traditions connected with the Prophet. This domain of investigation did not previously exist and is exclusively developed from scratch by the Muhadditheen. No such scientific research domain exists in European sciences, hence a lack of terminology for it. Europeans did not learn this science. They have, therefore, not employed it in their study of disciplines like gospel, history, or in the modern age to weed out spurious phenomena like the fake news or alternate facts.

There is no counterpart of Factology in English language, nor in Christian traditions.

Muhadditheen

The science of Hadith is chiefly an application of factology. A muhaddith collects ahadith (plural of hadith) by seeking out people who have a hadith to narrate. The narrator is known as the Ravi. It is incumbent upon a ravi to satisfactorily explain how he or she received the narration; starting from the source and declaring each and every intermediary ravi between the source and himself or herself. The factologist uses a most detailed and expansive catalogue of all the ravis that exist in the total landscape of narrations. A detailed biography is compiled about each ravi and it includes factors like piety, truthfulness, profession, etc. Any doubtful character among the chain of ravis for a narration will make the narration fail the test of factology. A great body of knowledge exists in the science of factology, for example the evaluation of the piety of each ravi by reviewers over the history. History thus is not just a collection of assertions by historians, rather each such assertion is fact checked most thoroughly according to the most strict and well specified criteria of factology. Therefore, Muslim muhadditheen (plural of muhaddith) are historians with far greater reliability than the European historians who can maneuver with respect to accuracy, because no comparably rigorous fact checks are applied to their works.

It is incumbent on the muhaddith in question who directly receives a rivayet (a narration as it was narrated by its ravi) from a ravi to fact check that ravi with respect to his or her truthfulness, piety, and other details of character. A muhaddith who fails to properly fact check his sources and the links in the chains of a rivayet is not reput-able, and his ahadith are not acceptable.

There is no counterpart of Muhaddith and Factology in Christianity, nor indeed in the Western idea of history as the system of peer review and citations is at best skeletal in comparison, and therefore no corresponding terms exist in related languages.

Qiyasists

The term represents scientists who investigate judicial inferences and precedences. It is developed chiefly by the people of the Fiqh. They analyzed Quran and compiled ahadith, incorporating precedence, to decide judiciously. They use direct qiyas (analogies) with respect to the material in precedents. Qiyasists try not to use their intellectual interpolation and extrapolation in making a decision, so the personal discretion of the qadhi is not involved; therefore, most qadhis would likely be onboard in commending such a decision.

Qiyas is somewhat similar to the practice of judicial precedence in constitutional judicial systems. There is, however, an important difference, namely, only those occurrences of precedence are incorporated that happened in early Islamic era and which have been verified and validated by the factologists.

Istihsanists

The term is used and developed by the people of the Fiqh. Scripture, compiled ahadith, and precedents are used, as is the case with a qiyasist. Application of qiyas solves many of the judicial cases. However, in some judicial cases the application of qiyas does not lead to a unique decision, because in such cases the processes of qiyas do not converge. It can happen for many reasons. First, there may exist no precedence because the world conditions have changed in terms of the possible set of actions and events. An example would be the performance of rituals like prayer, fasting, and hajj during air travel. Second, the application of qiyas may not yield a clear and unique judgement because the set of qiyas processes available in a judicial case and related circumstances may be varied and diverse. In general, when the application of qiyas is insufficient to arrive at a unique judicial decision, the Qadhi needs to resort to the use of his own intellect to arrive at an istihsan with respect to what the correct judgement should be in a particular situation. In this process, the personal discretion of the qadhi is involved in such a way that benefit of doubt is amply allowed in favor of the accused. This is the essence of 'istihsan'.

In the case of qiyas the use of intellect is not specific to the particular qadhi since most qadhis would be onboard. However, that is generally not the case in the exercise of istihsan, so that the istihsan-based judgement may be more specific to the particular qadhi. Such distinction and finetuning points to the sophistication of Islamic judicial system. No such distinction and corresponding sophistication exists regarding the use of precedence among the European justice systems, and that can and does lead to arbitrariness and systematic bias.

Mufassireen

The Quran was revealed among a particular segment of humanity and at a particular historical epoch in the world. As Islam reached extended regions of the world among different segments of humanity, and the time flowed on to new epochs in human history, it became necessary to reevaluate the message of Quran for these new segments of humanity and for the new epochs in human flow. This is an obvious need not just for the Arabs, who were the original addressees of the Quran, but also internationally and globally. It is a very divisive, and sometimes even violent, question among Muslims as to how this task shall be performed.

Generally, there are two processes to do this task. They use what is called Tafsir and Ta'wil. While all Muslims acknowledge the need for Tafsir, a significant segment among Muslims does not acknowledge the need for Ta'wil, and some will even question if it makes sense to do Ta'wil. This circumstance exists notwithstanding the fact that Muslims do not agree upon definitions and scopes of these two processes. Some of the confusion is created by the Orientalists who use limiting and misleading terms in related discourses. For example, they generally use the term 'traditionalist' for people of and practitioners of hadith. Among many other issues, the use of the word 'traditionalist' produces much ambiguity and misinformation simply from the dictionary meaning of the word.

The process of Tafsir uses mainly the ahadith in order to explain the message of Quran. The process is sometimes limited by the requirement that any explanation of the Quran be compliant with, and to some extent subservient to, the work of Tafsir that has already been done by the past Mufassireen (people of Tafsir, with Mufassir its singular). This is especially so with respect to the Tafsir presented by the Ashab (apostils) of the Prophet, and those who immediately followed the Ashab, known as Tabi'in. However, the restrictive requirement is applied even for more recent works of Tafsir, that a coherent and persistent community of Muslims, known as Ulama, have decided to uphold with their approval.

The circumstance makes the task of doing Tafsir insurmountably resistant for a newcomer. And we have not even begun to talk about Ta'wil about which there is not even a faint agreement among Muslims.

Awwalagists

Given the situation with Tafsir and Ta'wil, we have decided to use a fresh concept that we have called Awwalogy. The term Awwalagists represents scientists who perform analysis to bring the textual meanings back to their roots, the original meaning intended by the text.

The term is intended to include both tafsir and ta'wil, as well as additional research. Tafsir means to explain, to expound, to elucidate, and to interpret. Ta'wil means to return, to revert, thereby returning to the original meaning of a word to understand its connotations.

This is not the same as exegetes because the term exegesis applies heavily to the Bible paying little heed to the process for other scriptures and languages. Also, the term exegetes is not rich enough to discriminate between tafsir and ta'wil, for instance. Further, the term exegesis is used also in broad connotations such as politics and literature. Therefore, the term Awwalagists is coined to represent the concept of research in Ta'wil and Tafsir.

The term Awwalagist carries a meaning that is fresh even for Muslim tradition. It is distinct from the term mufassir which carries quite a historical baggage and is entrenched in sectarian controversy. Independent of such historical baggage and controversy, the term Awwalagist carries a spirit of research in pursuit of the truth, irrespective of who the researcher is in the conventional sectarian landscape.

There is no concept of Awwalogy in English language, nor in Christian religious tradition where the integrity of religion and its continuity is delegated to Ecumenical Councils and the institute of Papacy.

Fiqh

Ideally, a mujtahid should specify fiqh, which is the set of rules for judicial decision making, and a set of such decisions. However, the five Imams who ventured into specifying the five fiqhs, were not Mujtahids. Therefore, these fiqhs are not ideal. Consequently, they differ among themselves sufficiently to warrant five different shades of Islam.

Faqih

A faqih is one who can use the rules and decisions set of fiqh to make judicial decisions in a living court. The plural of faqih is fuqaha.

Qadhi

A qadhi is a faqih who is officially appointed to make binding judicial decisions that are enforced under the state authority.

In this volume, we use the terms Fiqh, faqih (whose plural is fuqaha), and qadhi. Many translators have used jurisprudent, jurist, and judge in their place; however, the reader should be aware that they are not equivalent.

Kalam

Kalam is the science of research into nature of Allah, significance of the Names of Allah, significance of Quran and other scriptures, significance of Prophet Mohammad and other prophets, the significance of the angles, and the meaning and significance of the articles of Iman. As a result of discourses in Kalam, a rational concept emerges regarding the desirable attitude towards Allah and the Prophet in life, and the meaning and significance of Iman. This concept is called 'Aqeeda. As is clear from this discussion, 'Aqeeda is vaster than Iman as Iman is only a component of 'Aqeeda. Iman is described in Quran as commitment to the significance and role of Allah, Prophet, Scriptures, Angels, and the Resurrection on the Day of Judgement. Whereas, 'Aqeeda is not discussed in the Quran as a distinct concept.

A person who is skilled in Kalam is called Mutakallim whose plural is Mutakallimeen. We will use Kalam to represent the philosophical content of the religion; and we will use Mutakallim for the practitioner of kalam.

These terms are not equivalent to the Christian terms, theology and theologian respectively.

The plural of mutakallim is mutakallimeen.

Ruhaniyaat

Spirit and soul are not Islamic; they are Christian concepts. Their use by the translators, are sometimes misleading due to their Christian content and baggage.

The one word that Quran uses is Ruh, which is sometimes translated as the "breath". That again is simple minded on the subject of ruh, for Quran itself declares that it is a secret among the secrets.

Any experience of the ruh is referred to as Ruhani experience. The act of seeking such an experience is called Ruhaniyat. Its plural is Ruhaniyaat, which also means the science of ruh.

The term represents the science of Islamic seeking of the truth about life, generally based on the practices of Salat (prayer), Saum (fasting) and Zikr (recall). These people are Scientists of Ruhaniyyaat. Non-Muslim investigators like the Christian Saints might also be candidates if they meet the criteria. However, Muslims enjoy far greater freedom from the central religious authority, which does not exist in Islam or among Muslims, who practice ruhaniyyaat freely, without coercion from a central authority.

Muslim "seekers" have therefore been able to make ruhaniyyaat into a genuine science with detailed methodologies and verification markers along the path, technically called the Tariqa.

On the other hand, Christians generally think that it is from Grace and not from Deeds, and thus the path (Tariqah) becomes almost irrelevant for Christians. Further, the Church confers Sainthood by its own initiative and criteria. And it happens after the death of the candidate. This is so unlike a science, so that Christian Saints do not generally qualify as seekers of Ruhaniyyaat, or Sufis.

Haqiqah

The goal of ruhaniyyaat is to know the Haqiqah which signifies the reality of life. There is a nominal way to proceed in search of Haqiqah, and this way is called *Tariqa.* One essential part of Tariqa is to purify the inner self of one's own self. This process is known as *Tazkiyah.* One way for Tazkiyah is to practice *Zuhd*, which means, in part, that the seeker avoids luxury in life and adopts a type of detachment from the worldly aspects of life in order to establish a strong attachment with Allah: while of course adhering to Salat, Saum and Zakat as integral to Ruhaniyyaat.

Sufiism

Quran declares that ruh is a decision among the "Decisions" of God, and what knowledge of it can be given to people is but a little bit.

Every one concentrates on Salat, Saum and Zakat. For most people this practice is an end in itself, because they feel this practice can afford them salvation in the eyes of God. Sufis also concentrate on the same practices. The Sufis, however, tend to use the practices of Saum, Salat, and Zakat as a means to know the "Decisions of God", even if only a little bit. Doing so sets the Sufis on a different but parallel and equivalent path to most Muslims who make Salat, Saum and Zakat as goals within themselves. It is sometimes stated that the Sufi approach allows a deeper experience with the fundamentals, namely the practice of Salat, Saum and Zakat.

Sufiism is a science the counterpart of which does not exist in English as a language, nor in Christianity as a religion. Even though Christianity has Saints, they are not equivalent to Sufis for two main reasons. First, no science like Sufiism exists among Christian saints, for example, no Tariqah is rationally devised for Sainthood: on the contrary, some Christians profess that it is by grace, not by deeds. Second, the Saintship is awarded, posthumously, by the Church on its own initiative and discretion.

Mujtahid

A mujtahid is someone who can use ijtihad, which roughly means exercising qiyas and istihsan in an error-free way, and can pass judgement on the matters of other non-fiqh religious affairs in an error-free way. A mujtahid is knowledgeable about the scriptures, their tafsir and ta'wil, the application of qiyas, and limitations of istihsan, and practice of awwalogy, factology, and hadith, etc. In addition, a mujtahid conforms to the scriptures in an exemplary way, and is also exemplary in his inter personal dealings with people. He or she is also an established qiyasist, istihnasist, factologist, muhaddith, and awwalagist, etc. A mujtahid is one who can exercise ijtihad, which is a process that exercises all the above-mentioned elements with exceptional accuracy, almost error free.

A mujtahid exercises all Islamic knowledge to accurately extend it to situations not initially experienced or anticipated. There is no equivalent word for Mujtahid in English language.

Qiyas and istihsan are necessities in fiqh, and they are not binding outside the legal system. While all fuqaha and qadhis can exercise qiyas and istihsaan, directly or indirectly via consultation with other qadhis, they are generally not qualified to exercise ijtihad for Muslims in general.

Muslims are not all onboard about the scope and permissibility of using qiyas, istihsan, and ijtihad.

Five Imams are popular: Imam Abu Hanifa, Imam Malik, Imam Shafi, Imam Hanbal, and Imam Jaffer Sadiq. However, this epithet "Imam" is only in a lose sense of popular public approval. Their exercise of qiyas and istihsan and other ulum in Islam was not generally accepted widely in learned circles: and that is why we have five of them, and they were almost contemporaries, living over a short period of time. The general Muslim ulama became so concerned with the situation that they refused to welcome any more Imams in fiqh, insisting that five Imams are enough and all people should follow one or another of them. That gave rise to the practice of Taqlid among those who confess to be ahl al-sunnah wal jama'h, which caused deterrence to independent research and scholarship among them.

If the five Imams are not Mujtahid, and the door of Ijtehad is closed by the practice of Taqlid, who then is a Mujtahid among them?

English terms for some of the other sciences are retained for the Scientists of the Islamic Era, because such terms are approximate enough. For example, mathematics, physics, surgeon, sociology, economics, political science, poetry, philology, and religion etc. Even though Muslims often approach these topics with subtly different angles, for now we will decide to use these terms in the context of Scientists of the Islamic Era Series. However, a need still exists to introduce additional terms in order to represent the research of the scientists of the Islamic Era at a finer resolution. We will leave that for future editions.

It is also necessary to remark how the historical periods are currently named and defined. These are currently European centric; as well as confusing and lacking clear rational to the terminology. We have corrected the anomaly at a small scale, namely as it concerns the definition of the "Middle Ages". These are also variously called Dark Ages or Medieval Period (also written as Mediaeval Period). The period has vague start and stop durations, as treated by different historians. We have, therefore, redefined it crisply, and also renamed it. Instead of starting it from 476 AD, a vague reference point to the fall of Roman Empire, we start it at 610 AD, a reference point to the beginning of the Islamic Era when the Qur'anic revelation began. We end this Islamic Era with the fall of Ottoman

Caliphate in 1922. Thus, the Middle Ages is largely replaced with the Islamic Era, and it extends into what is commonly and non-transparently called the Modern Era, Modern Age or Modern Period. Doing so, makes the scheme less European centric and more logically consistent.

We present this series of eight volumes to the readers to share with them the wealth of scientific excellence that these scientists contributed towards greatly extending the frontiers of all these sciences. In particular, they enhanced the world civilization towards prosperity, democracy, justice and peace; as well as produced Renaissance in Europe. The series also aims to bring awareness to the Muslim readers about their role as the torch bearers of science and civilization, and to serve the upwelling thirst that the young generation have for the truth about Islamic civilization. The series urges the academicians and researchers of the world, especially the Europeans and Americans, to learn and celebrate the Muslim giants of science upon whose shoulders they stand, and without whom the present-day scientific achievements could not have been possible.

We have not explicitly stated the inclusion criterion for the scientists in the series, "Scientists of the Islamic Era". We state it briefly here. We have included only those scientists whose historicity is validated, those who have explicitly contributed through their writings and their works, and whose biography falls within the declared time period of focus in this series. Such criteria would not let us include mythical entries, and those without a trail of writings and works.

Researchers like Professor Fuat Sezgin have, at great length, researched and investigated contributions of the Muslim scientists. He has edited 1600 volumes. Such work is invaluable for projects like ours. Some of this work is summarized online, such as in Wikipedia articles available under GNU free document license.

We aim to serve our community, inspire them and our coming generations, and inform them of their role as torch bearers of excellence in world science, technology, and civilization.

The Muslim scientists lived an integrated life with no conflict between their religion and their scientific passions; and, unlike the many present-day scientists, a question never occurred that their scientific passion somehow needed to be separate from their religious inspirations. This is also obvious from the fact that most scientists were themselves experts in Islamic jurisprudence, hadith and Quran. In reality, their scientific work was also their religious worship because Islam showed them the necessity to do science and motivated them for it. Islam encouraged their scientific passion by equating it with religious worship. No wonder they achieved scientific excellence with amazing integrity, generosity and grace.

The Muslim education system was very different from what the world has today; and judging from the results, it was greatly more affective and better integrated into overall life. The biographical summaries will provide tiny glimpses into this system. It avoided narrow specializations and produced polymaths. We have Muslim scientists who are simultaneously excellent Quranic scholars, jurisprudents, mathematicians, astronomers, physicists, medical professionals, chemists, botanists, physiologists, poets, men of letters, and grammarians, etc.

To keep this series on "Scientists of the Islamic Era" brief we have included each scientist only once, under a category we deemed appropriate, given their works and biography. As a consequence, for instance, not all natural scientists appear in volume 1 which is dedicated to the Natural Scientists, because we decided to include some in Volume 2 which is dedicated to the Medical Scientists; similarly, not all medical scientists appear in Volume 2 because we decided to include some in Volume 1. This situation invariably occurs across the board. Another consequence is that, for instance, if a scientist is included in Volume 2 for the Medical Scientists, his or her achievements in Astronomy, for instance, are not highlighted.

For such reasons, the preface to the series will get edited as the series progresses and situations for it arise.

The scientists are listed in chronological order, allowing an opportunity to correlate scientific tides and ebbs with political and religious ups and downs.

They could have been ordered according to the significance of their scientific contributions; that, however, is problematic because it is difficult, if not impossible, to assess the importance of research and compare across different scientific disciplines within sciences.

The order could have been sequenced according to how well the scientists are known today; that too is problematic because not all excellent scientists are well-known today, and, those who are, generally are made famous by the European commentators, who often did not know their works in original Arabic, and did not reflect the actual significance of their research. The well-known-ness is fairly arbitrary. For instance, Omar Khayyam is celebrated today for his Rubaiyat, which was something he did on the side, while his real works were in mathematics, a fact that is largely obscured.

This series of books should add to the impulse that is now thrusting the Muslims into the world of science and technology with increasing excellence in their achievements, signaling that their own Renaissance has now renewed.

Muslim Voice
Bowie, MD, USA.
March 24, 2025.

Preface to the First Edition of Volume 4: Religious Scientists

"Scientist of the Islamic Era" is a book series encompassing eight volumes; the present book is volume 4 titled "Religious Scientists of the Islamic Era". It covers 84 religious Scientists encompassing people of tafsir and ta'wil, ilm-al-kalam, hadith, fiqh, justice, ijtihad, and ruhaniyyaat. The period of coverage is part 1 of the Islamic Era, from AD 610 to 1400. They commanded exceptional breadth in their learning and deepest insights in their specializations. They formulated the foundations of the fields of knowledge in the Religious Sciences and defined its frontiers. This field is unique to Muslims and Islam because Islam is a religion free from dogmas and therefore amenable to scientific treatment, and consequently Muslims are the ones who were free to develop the required scientific world-view with which to formulate the related sciences.

Each scientist is briefly described. First, the name of the scientist is disambiguated, and an attempt is made to correct the misrepresentations common in the European translations. Salient scientific contributions of each scientist are briefly highlighted, a difficult task because of the fact that many of these scientists were polymaths. For each scientist we have provided a biographical summary to help picture their love and craving for knowledge, and the motivations and opportunities for them to do their research.

It is our objective that this fourth volume in the series will inform the Muslims about the wealth of their scientific heritage, and the next generations will feel inspired to surpass the excellence of their ancestors to enrich their heritage further, and be, like their ancestors, the flag bearers of world civilization in the religious sciences.

Muslims are now excelling in research with superb agility, and this series on Scientists of the Islamic Era will further stimulate this ongoing Renaissance in the Muslim world.

Abdur Rahim Choudhary, Ph.D.
Bowie, Maryland, USA
arc@muslimvoice.org
March 24, 2025.

Religious Scientists

Religious Sciences community in Islamic Era was a unique phenomenon. It brought to humanity people who seamlessly integrated sciences and religion. Their religion was new in the world, and their passion for sciences was unparalleled. The religion was free from dogmas and therefore was already based on logic and scientific principles. These principles included the central importance of observation, analysis and theorization based on facts, not dogmas.

Europeans learned the research discoveries of Muslim Scientists, and adopted this research as their own, without acknowledging the source. The Europeans pretended that these research achievements were their own, without indebtedness to the Muslims. Thus, astronomy was started with Kepler et al, the calculation of orbits of the celestial bodies started with Newton et al, pharmaceuticals were discovered by Pasteur et al, and philosophy was founded by Descartes et al, etc. All this happened at such a fast pace that had never been witnessed before. It was super intelligent Europeans who had spent centuries in the Dark Ages: they came out of darkness and started championing scientific discoveries that their forefathers could not even imagine. And the European mind does not even wonder at this discontinuity, and in cases when it does, it finds comforting caresses in Greek sciences, which were more than a millennium out of date, and limited in their scope, foundations and frontiers.

The answer is that they learned, without acknowledgement, the superb scientific and civilizational achievements from Muslims. They experienced renaissance in all the sciences that the Muslims had invented, except one. The exception was the Religious Sciences. The reason was the non-logical and patchy foundation of Christianity that refused to come under the systematics of a science. The result was a breakup of Christianity into Catholics and Protestants. Even the protestants could not give their religion a non-dogmatic structure. The result is that religious sciences in Europe and its extensions are non existent, though there are plenty of churches, seminaries, monasteries, cathedrals, priests and their long hierarchy, and even saints as canonized by the church.

These facts are obvious even if one examines not the entire scientific works by the Muslim scientists but only a subset of those that had been very visibly translated into European languages.

This book describes 84 religious scientists from part 1 (610-1400) of the Islamic era covering the disciplines of ilm-al-kalam for a scientific discourse on the scriptures; factology for a scientific way to fact check the narrations and historical assertions; hadith for a compilation of facts about the way of the prophet Mohammad, using the science of factology; fiqh to lay down an elaborate structure to formulate, judge and serve justice, based on Quran as the constitution, and Hadith as a compilation of precedences that were fact checked by the factologists; qiyas and istihsan frameworks to allow for logical analogies in serving justice by the faqih; ruhaniyyaat to explore the deeper role, meaning and scope of the foundational practices of Salat, Saum, Zakat and Hajj; Sufis as the people who find Tariqas (paths) to systematically seek ruhaniyyaat; and ijtihad to encounter situations that were not experienced in the time of the Prophet, or cannot be reduced to such using qiyas and istihsan. There are also other areas of religious sciences.

Each scientist is briefly described. First, the name of the scientist is disambiguated, and an attempt is made to correct the misrepresentations all too common in the European translations. Salient scientific contributions of each scientist are briefly highlighted, a difficult task because of the fact that most of these scientists were polymaths. For each scientist we have provided a biographical summary to help picture their love and craving for knowledge and the motivations and opportunities available for them to do their research.

The 84 religious scientists are organized in the following seven categories: 10 Mohadditheen, 25 Fuqaha and Qadhis, 6 Mutakallimeen, 4 Awwalagists, 12 Fuqaha with their own Fiqh, 26 scientists of Ruhaniyyaat, and 1 Mujtahid.

A brief description for each scientist is provided, each in a separate subchapter. The 84 subchapters, in this book, are each dedicated to a single religious scientist. Some chapters are short, while others are detailed. Information on these topics is not abundant because the existing research is at best sporadic, and is mostly championed by individuals or small groups. There is a strong need for more detailed studies, on sustained and institutional bases, on an expansive scale.

The present series of eight volumes is offered in this context. They are intended for the educational and research institutes, at national and international levels, to provide encouragement for further work focused along these lines.

Mohadditheen

Quite often English writers and translators use the term "traditionalist" which is not clear and transparent, and causes non clarity and confusion. We have introduced the Islamic terms to be used instead. These are well-defined, unambiguous, and transparent in their meanings, causing no confusion.

The first such term is **factologists**. It represents the scientists who investigate criteria used for the verification and validation of orally transmitted traditions or written historical statements. The science of factology was developed and practiced by the Muhadditheen, people who collect ahadith from the ravis. The science, among other things, uses what is known as 'knowledge about the people (Ilm-ur-Rijal) encompassing all people who narrated any hadith connected with the Prophet. This domain of investigation did not previously exist and is exclusively developed from scratch by the Muhadditheen. No such scientific research domain exists in European sciences, hence a lack of terminology for it. Europeans did not learn this science. They have, therefore, not employed it in their study of disciplines like gospel, history, or in the modern age to weed out spurious phenomena like the fake news or alternate facts.

There is no counterpart of Factology in in Christianity as a religion, nor in English as a language.

A **mohaddith** applies factology to individually collected ahadith, and to the chain of ravis in each hadith. A muhaddith collects ahadith by seeking out people who have a hadith to narrate. The narrator is known as the ravi. It is incumbent upon a ravi to satisfactorily explain how he or she received the narration; starting from the source and declaring each and every intermediary ravi between the source and himself or herself. The factologist uses a most detailed and expansive catalogue of all the ravis that exist in the total landscape of narrations. A detailed biography is compiled about each ravi and it includes factors like piety, truthfulness, profession, etc. Any doubtful character among the chain of ravis for a narration will make the narration fail the test of factology. A great body of knowledge exists in the science of factology, for example the evaluation of the piety of each ravi by reviewers over the history. History thus is not just a collection of assertions by historians, rather each such assertion is fact checked most thoroughly according to most strict and well specified criteria of factology. Therefore, Muslim muhaddith are historians with far greater reliability than the European historians who can maneuver with respect to accuracy because no rigorous fact checks are applied to their works.

The plural of muhaddith is muhadditheen.

It is incumbent on the muhaddith in question who directly receives a narration from a ravi to fact check that ravi with respect to his or her truthfulness, piety, and other details of character. A muhaddith who fails to properly fact check his sources and the links in the chains of narrations is not reputable, and such ahadith are not acceptable.

There is no counterpart of Muhaddith and Factology in in Christianity as a religion, nor in English as a language.

1. Sufyan ibn `Uyaynah

Abū Muḥammad Sufyān ibn ʿUyaynah ibn Maymūn al-Hilālī al-Kūfī (Arabic: ابو محمد سفيان بن عيينة بن ميمون الهلالي الكوفي),

(725 – February 25, 814),

was a prominent eighth-century alim from Mecca. He was from the third generation of Islam referred to as the Tābiʿu al-Tābiʿīn, "the followers of the followers". He specialized in the field of hadith and Qur'an tafsir and was described by al-Dhahabī as shaykh al-Islam - a preeminent Islamic authority. Some of his students achieved much renown in their own right, establishing schools of thought that have survived until the present.

Scientific Contributions

Ibn ʿUyaynah compiled one of the early collections of hadith; his Jāmiʿ followed the Muwaṭṭa' of Mālik ibn Anas. The subject of his book was Prophetic narrations (sunan) and subsequent narrations (āthār). He also wrote "tafsir" of Qur'an.

Al-Ubbī, a latter religious scholar, claimed this work to be one of the first compilations in Islam. In summary, his two known works are:

- al-Jāmi'
- al-Tafsīr

Ibn ʿUyaynah was praised by contemporaries for both his knowledge and humility. ʻAbd al-Raḥmān ibn Mahdī described him as from the most knowledgeable people of the hadith of the inhabitants of the Tihamah region of what is now Saudi Arabia.

He was lauded by Muḥammad Ibn Ismā'īl al-Bukhārī for his memorizing ability, an essential quality for a hadith narrator.

Not just a transmitter of recorded knowledge, his student al-Shāfiʿī said he had not seen anyone more adept at explaining the meaning of hadiths than Ibn ʿUyaynah.

His humility was also illustrated by al-Shāfiʿī's mention of Ibn ʿUyaynah's reluctance to give religious verdicts (fatwas).

Ibn Mahdī preferred him to a contemporary of his, Sufyān al-Thawrī, in their understanding of the Qur'an and hadith.

Statements attributed to Ibn ʿUyaynah illustrate his respect for religious knowledge, acting upon that knowledge and the sacrifice necessary to obtain it. In one statement he said that whatever increase a person experience in intellect, is matched by a decrease in their material wealth. In another statement he stated that knowledge that does not benefit an individual is of detriment to them.

Ibn ʿUyaynah's students were numerous. Many of them would embark on a Hajj (pilgrimage) to Mecca intending to meet him and then crowding him during the days of Hajj. Some of Ibn ʿUyaynah's teachers were also his students, for example, al-ʿAmash, Ibn Jurayj, and Shu'bah. Both Abū ʿAbdullāh Muhammad ibn Idrīs al-Shāfiʿī, the namesake of the Shāfiʿī school of fiqh, and Aḥmad ibn Muḥammad ibn Ḥanbal, the namesake of the Ḥanbalī school, were his students.

Al-Nawawī, a prominent Shāfiʿī scholar, cited Ibn ʿUyaynah as from "the grandfathers of the Shāfiʿī scholars in their methodology in fiqh". The ahadith Ibn ʿUyaynah narrated were later included in the six canonical hadith collections.

Biographical Summary

Ibn ʿUyaynah's father, ʿUyaynah ibn Abī ʿImrān, was originally from Kufa in present day Iraq where he was a governor for Khālid ibn ʿAbdillāh al-

Qasrī. However, when al-Qasrī was removed from his position 'Uyaynah fled to Mecca where he then settled.

Ibn 'Uyaynah was born in the year 725 CE/107 AH and began his religious studies while still young. He said of himself that he first sat formally with a religious instructor at 12 when he attended the lessons of 'Abd al-Karīm Abū Umayyah. Subsequent teachers include 'Amr ibn Dīnār, al-Zuhrī, Ziyād ibn 'Allāqah, Abū Isḥāq, al-Aswad ibn Qays, Zayd ibn Aslam, 'Abdullāh ibn Dīnār, Manṣūr ibn al-Mu'tamir, 'Abd al-Raḥmān ibn al-Qāsim and many others.

He was the client (mawlā) of Muḥammad ibn Muzāḥim.

By his own account, Ibn 'Uyaynah read the entire Qur'an (perhaps meaning that he had memorized it) by the age of four and began writing hadith at age seven. Upon turning 15, his father gave him the following advice which he later said he never turned away from:

> My son, the meanderings of childhood have now departed you, associate yourself with good and you will be from its people. And, know that none will be content with the religious scholars unless obedient to them so obey them and be content, serve them and grasp some of their knowledge.

He lived in Mecca and had nine brothers. Of the brothers, five pursued studies in hadith with Sufyān becoming the most renowned of them. The names of the remaining four are Muḥammad, Ibrāhīm, Ādam and 'Imrām.

Ibn 'Uyaynah died in 814 AD.

Ibn 'Uyaynah performed the Hajj (pilgrimage) seventy times, saying that each time he went he supplicated Allah that that not be the last time he visited the places of Hajj. He said he was shy to ask this again on the seventieth occasion and returned to Mecca and died there within the next year. He died on Saturday February 25, 814 CE, the first day of Rajab, 198 AH, at the age of 91. He was buried in the al-Ḥajūn district of Mecca.

2. Al-Darimi

Abu Muhammad 'Abdullāh Bin 'Abd Ar-Rahman Bin Fadl Bin Bahrān Bin 'Abd As-Samad At-Tamīmi Ad-Dārimi As-Samarqandī

(Arabic: الدارمي) (181–255 AH / 797–869 CE),

was a mohaddith of Arab or Persian background. His best known work is Sunan al-Darimi, a book collection of hadith.

Scientific Contributions

Al-Darimi transmitted hadiths from Yazid ibn Harun, Abd Allah ibn Awn, and others. A number of mohadditheen transmitted hadiths from him, including Muslim ibn al-Hajjaj, Abu Dawood, Al-Tirmidhi, and Abu Zur'a al-Razi.

Following are some works by Al-Darimi.

- Sunan al-Darimi - Some from among his collections of the Prophet Muhammad's ahadith.
- Tafsir al-Darimi - Imam Dhahabi mentioned the work in Siyar A'lam al-Nubala. Not extant
- Al-Jami'a - Khatib al-Baghdadi has mentioned this in his Ta'rikh al-Baghdad.

Biographical Summary

Imam Darimi, came from the family tribe of Banu Darim Bin Maalik Bin Hanzalah Bin Zaid Bin Manah Bin Tamim or Banu Tamim the Arab tribe from Arabian Peninsula. He is also known as Imam Tamimi, in relation to Tamim Bin Murrah, who was amongst the ancestor of Banu Darim.

As stated by Darimi "I was born on the same year in which Imam Abdullah Bin Mubarak had died. And Abdullah Bin Mubarak died in 181 AH". He was born in Samarkand in 797 AD, and he died in Muscat in 869 AD.

3. Al-Tabarani

Abu 'l-Qāwsim Sulaymān Ibn Ahmad ibn Ayyoob ibn Muṭawyyir al-Lakhmī ash-Shāmī at-Ṭabarāni

(260 AH/c. 874 CE - 360 AH/971 CE),

was a mohaddith and a faqih.

Scientific Contributions

He is known primarily for three works on hadith:

- al-Muʿjam al-Kabīr – from which he excluded the traditions of Abu Hurayra
- Al-Mu'jam al-Awsat – which contains traditions from Abu Hurayra
- Al-Mu'jam as-Saghir – which gave a hadith from each of his masters.

From amongst his students were: Ahmad bin 'Amr bin 'Abdul-Khaliq Al-Basri and Abu Bakr Al-Bazzar.

Biographical Summary

Imam Al Tabarani was born in 260H (873 AD) in Tabariya As-Sham. He narrated Hadiths from more than one thousand ravis. He traveled extensively to many regions to quench his thirst for ahadith. He traveled to Syria, Haramayn, Tayyibayn, Yemen, Egypt, Baghdad, Kufa, Basra and Isfahan etc.

He wrote many Hadith books, among them are Al-Mu'jam Al-Kabir, Al-Mu'jam Al-Awsat, and Al-Mu'jam As-Saghir.

Abul 'Abbas Ahmad Bin Mansoor states: I have narrated three hundred thousand Ahadees from Imam Tabarani. He lived most of his end life in Isfahan, Iran, and died there on 27th Dhul-Qa'da, 360 H (970 AD).

4. Muhammad al-Bukhari

Abū ʿAbd Allāh Muḥammad ibn Ismāʿīl ibn Ibrāhīm al-Juʿfī al-Bukhārī

(for short Muhammad al-Bukhari),

(Arabic: أبو عبد الله محمد بن إسماعيل بن إبرهيم الجعفي البخاري),

(21 July 810 – 1 September 870),

was a 9th-century *muhaddith* who is widely regarded as the most important mohaddith and factologist. He is well known for his hadith collection called Sahih al-Bukhari.

Scientific Contributions

Sahih al-Bukhari is revered as an important *hadith* collection in Islam. *Sahih al-Bukhari* and *Sahih al-Muslim,* the *hadith* collection of Muslim ibn al-Hajjaj, are together known as the Sahihayn (Arabic: صحيحين) and are regarded as authentic books of hadith. It is part of the Kutub Sihah al-Sittah, the six Sahih (Correct) collections of *hadith.*

According to hadith scholar and historian Al-Dhahabi, al-Bukhari began studying hadith in the year 821 CE. He memorized the works of Abd Allah ibn al-Mubarak while still a child and began writing and narrating hadith while still an adolescent. In the year 826 CE, at the age of sixteen, Al-Bukhari performed the *Hajj* with his elder brother and widowed mother. Al-Bukhari stayed in Mecca for two years, before moving to Medina where he wrote *Qadhāyas-Sahābah wa at-Tābi'īn,* a book about the companions of Muhammad and the *tabi'un.* He also wrote *Al-Tārīkh al-Kabīr* during his time in Medina.

Al-Bukhari is known to have travelled to most of the important Islamic learning centers of his time, including Syria, Kufa, Basra, Egypt, Yemen, and Baghdad. He studied under prominent Islamic scholars including Ahmad ibn Hanbal, Ali ibn al-Madini, Yahya ibn Ma'in and Ishaq ibn

Rahwayh. Al-Bukhari is known to have memorized many hundred thousand *hadith* narrations.

Al-Bukhari was driven out of Nishapur. Al-Dhahabi and al-Subki asserted that Al-Bukhari was expelled due to the jealousy of certain ulama of Nishapur. Al-Bukhari spent the last twenty-four years of his life teaching the *hadith* he had collected. During the *mihna*, he fled to Khartank, a village near Samarkand, where he then also died on Friday, 1 September 870.

Sahih al-Bukhari is considered Al-Bukhari's *magnum opus*. It is a collection of approximately 7,563 *hadith* narrations across 97 chapters creating a basis for a complete system of fiqh without the use of qiyas. The book is highly regarded among Muslims, and most ulama consider it authentic.

Al-Bukhari wrote three works discussing narrators of hadith with respect to their ability in conveying their material. These are *Al-Tārīkh al-Kabīr*, *Al-Tarīkh al-Awsaṭ*, and *Al-Tarīkh al-Ṣaghīr*. Of these, *Al-Tārīkh al-Kabīr* is published and well-known, while Al-Tarīkh al-Ṣaghīr is lost. Al-Dhahabi quotes Al-Bukhari as having said, "When I turned eighteen years old, I began writing about the companions and the *tabi'un* and their statements. At that time I also authored a book of history at the grave of the Prophet at night during a full moon." The books being referred to here were *Qadhāyas-Sahābah wa at-Tābi'īn* and *Al-Tārīkh al-Kabīr*. Al-Bukhari also wrote al-Kunā on patronymics, and Al-Ḍu'afā al-Ṣaghīr on weak narrators of hadith. Al-Adab al-Mufrad is a collection of hadith narrations on ethics and manners.

In response to the accusations levied against him during his *mihna*, Al-Bukhari compiled the treatise *Khalq Af'āl al-'Ibād*, the earliest traditionalist representation of the position taken by Ahmad ibn Hanbal, in which Al-

Bukhari explains that the Quran is God's uncreated speech, while maintaining that God creates human actions, as the Sunnis had insisted in their attacks on the free-will position of Qadariyah. The first section of the book reports narrations from earlier scholars such as Sufyan al-Thawri that affirmed the Sunni doctrine of the uncreated nature of the Quran and condemned anyone who held the contrary position as a *Jahmi* or *Kāfir*. The second section asserts that the acts of men are created, relying on Qur'anic verses and reports from earlier hadirg ulama like Yahya ibn Sa'id al-Qatlan. In the last part of his treatise, Al-Bukhari harshly condemned the *Mutazilites*, defending the belief that sound of the Qur'an being recited is created. Al-Bukhari cited Ahmad Ibn Hanbal as evidence for his position.

A significant number of ulama, both historical and contemporary, maintain that al-Bukhari was an independent who did not adhere to any of the four famous madhhabs. Al-Dhahabi said that: Imam Bukhari was a scholar capable of making his own ijtihad without following any Islamic school of fiqh in particular.

Al Bukhari seems to allow tafsir and ta'wil versus a Zahirite approach. For example, al-Bukhari in his Sahih, in the book entitled "Tafsir al-Qur'an wa 'ibaratih", surat al-Qasas, verse 88, "kullu shay'in halikun illa Wajhah" [the literal meaning of which is "everything will perish except His Face"], al-Bukhari asserted the term [illa Wajhah] means "except His Sovereignty/Dominance". And, in this same chapter, there is another occasion when the term 'dahk' (Arabic: ضحك, lit. 'laughter') is narrated in a hadith, which is interpreted by "His Mercy".

Al-Bukhari rebuked those who rejected of *qadar* (predestination) in Sahih al-Bukhari; by quoting a verse of the Qur'an he implied that God had precisely determined all human acts. Al-Bukhari signified that, according to Ibn Hajar al-'Asqalani, if someone was to accept autonomy in creating his acts, he would be assumed to be playing God's role and so would subsequently be declared a *Mushrik;* similar to the later Ash'ari view of *kasb* (acquisition, occasionalism, and causality, which link human action with divine omnipotence).

Following are some works of al-Bukhari.

- Al-Tarikh al-Kabir.
- Kitāb al-Mukhtaṣar min al-tārīkh.
- Asāmī al-ṣaḥābah.
- Sahih al-Bukhari.
- al-Duʿafāʾ al-kabīr.
- Al-Duʿafāʾ.
- al-Duʿafāʾ al-ṣaghīr.
- Kitāb al-wuḥdān (On the Companions from whom only one hadith is transmitted) (not extant).
- Kitāb al-ʿilal (not extant).
- Birr al-wālidayn (hadith collection on filial piety).
- Al-Adab al-Mufrad.
- Kitāb al-hiba
- Al-Sunan fī al-fiqh = al-Fawāʾid = al-Mabṣūṭ (not extant).
- Al-Jāmiʾ al-Ṣaḥīḥ = al-Jāmiʿ al-kabīr = al-Musnad al-kabīr.
- Rafʿ al-yadayn fī al-ṣalāh.
- Al-Qirāʾa khalfa al-imām.

Biographical Summary

Muhammad al-Bukhari was born after the prayer on Friday, 21 July 810 (13 Shawwal 194 AH) in the city of Bukhara in Greater Khorasan in present-day Uzbekistan. His father was Ismail ibn Ibrahim, a scholar of hadith and a student of Malik ibn Anas, Abd Allah ibn al-Mubarak, and Hammad ibn Salamah. Ismail died while Al-Bukhari was an infant.

Al-Bukhari began learning *hadith* at a young age. He travelled across the Abbasid Caliphate and learned under several influential contemporary scholars. Bukhari memorized thousands of *hadith* narrations, compiling the *Sahih al-Bukhari* in 846 AD. He spent the rest of his life teaching the *hadith* he had collected.

Towards the end of his life, Bukhari was exiled from Nishapur. Subsequently, he moved to Khartank, near Samarkand. Today his tomb lies within the Imam Bukhari Mausoleum in Hartang, Uzbekistan; 25 kilometers from Samarkand. It was restored in 1998 after centuries of neglect and dilapidation. The mausoleum complex consists of Al-Bukhari's tomb, a mosque, a madrasa, library, and a small collection of copies of Quran. The modern ground-level mausoleum tombstone of Al-Bukhari is only a cenotaph, the actual grave lies within a small crypt below the structure.

5. Muslim ibn al-Hajjaj

Abū al-Ḥusayn ʻAsākir ad-Dīn Muslim ibn al-Ḥajjāj ibn Muslim ibn Ward ibn Kawshādh al-Qushayrī an-Naysābūrī

(Arabic: أبو الحسين عساكر الدين مسلم بن الحجاج بن مسلم بن وَرْد بن كوشاذ القشيري النيسابوري),

or Muslim Nayshāpūrī (Persian: مسلم نیشاپوری), commonly known as Imam Muslim,

(815 –875 CE / 206 - 261 AH),

was a muhaddith from the city of Nishapur (early Khorasan and present day Iran).

Muslim ibn al-Ḥajjāj's hadith collection, known as Sahih Muslim, is one of the six major hadith collections in Sunni Islam. It is regarded as one of the two most authentic (sahih) collections, alongside Sahih al-Bukhari.

Scientific Contributions

Muslim ibn al-Ḥajjāj's main contribution is "Sahih Muslim": his collection of authentic hadith. The Sunni scholar, Ishaq Ibn Rahwayh was first to recommend Muslim's work. Ishaq's contemporaries did not at first accept this; Abu Zurʻa al-Razi objected that Muslim had omitted too much material which Muslim himself recognized as authentic, and that he included transmitters who were weak.

Ibn Abi Hatim (d. 327/938) accepted Muslim as "trustworthy, one of the hadith masters with knowledge of hadith"; though this contrasts much more fulsome praise offered to Abu Zurʻa and also his father Abu Hatim.

Muslim's book gradually increased in stature such that it is considered among Sunni Muslims the most authentic collections of hadith, second only to Sahih Bukhari.

Biographical Summary

Muslim ibn al-Hajjaj was born in the town of Nishapur in the Abbasid province of Khorasan, in present-day northeastern Iran. Historians differ as to his date of birth, though it is usually given as 202-206 AH (819/822 AD).

Adh-Dhahabi said, "It is said that he was born in the year 204 AH," though he also said, "But I think he was born before that."

Ibn Khallikan could find no report of Muslim's date of birth, or age at death, by any of the hadith masters, except their agreement that he was born after 200 AH. Ibn Khallikan cites Ibn al-Salah, who cites Ibn al-Bayyi‘'s Kitab ‘Ulama al-Amsar, in the claim that Muslim was 55 years old when he died on 25 Rajab, 261 AH (May 875) and therefore his year of birth must have been 206 AH (821/822).

Ibn al-Bayyi‘ reports that he was buried in Nasarabad, a suburb of Nishapur.

According to scholars, he was of Arab ancestry. The nisbah of "al-Qushayri" signifies Muslim's belonging to the Arab tribe of Banu Qushayr, members of which migrated to the newly conquered Persian territory during the Rashidun Caliphate. Genealogists agreed Muslim was an Arab ethnicity hailed from Banu Qushayr. According to two other scholars, Ibn al-Athīr and Ibn al-Salāh, he was actually an Arab member of that tribe, of which his family had migrated to Iran, nearly two centuries earlier following the conquest.

Estimates on the number of hadiths in his books vary from 3,033 to 12,000, depending on whether duplicates are included, or only the text (isnad) is. His Sahih ("authentic") is said to share about 2000 hadiths with Bukhari's Sahih.

The author's teachers included Harmala ibn Yahya, Sa'id ibn Mansur, Abd-Allah ibn Maslamah al-Qa'nabi, al-Dhuhali, al-Bukhari, Ibn Ma'in,

Yahya ibn Yahya al-Nishaburi al-Tamimi, and others. Among his students were al-Tirmidhi, Ibn Abi Hatim al-Razi, and Ibn Khuzaymah, each of whom also wrote works on hadith. After his studies throughout the Arabian Peninsula, Egypt, Iraq and Syria, he settled in his hometown of Nishapur, where he met, and became a lifelong friend of al-Bukhari.

6. Ibn Abi Asim

Abu Bakr Ahmad bin `Amr ad-Dahhak bin Makhlad ash-Shaibani (Arabic: أبو بكر أحمد بن عمرو بن الضحاك بن مخلد الشيباني), widely known as Ibn Abi Asim (Arabic: ابن أبي عاصم), (822 – 900 AD), was a mohaddith of the 9th century from Iraq.

Scientific Contributions

Ibn Abi Asim compiled numerous Prophetic traditions into two volumes, organized into chapters based on different aqida-related topics. He had also written about the first-generation Muslim and Umayyad caliph, Mu'awiyah, though the work is now lost.

Regarding the topic of this book Al-Suyuti says it was a book on Mu'awiyah's dreams, while Ibn Hajar referred to it as a book on Mu'awiyah's virtues. Ibn Abi Asim's essay may have included both topics.

Historians Abu al-Abbas al-Niswi and Abu Nu`aym both reported Ibn Abi Asim as having been a Zahirite. Although he has become an important figure for the Zahiri school in the modern day, few of his works in fiqh have survived to the modern era.

Biographical Summary

Ibn Abi Asim was born in Basra, Iraq in 822. He grew up in an academic household, as both his father and his grandfather were mohadditheen in their own right. Due to his family's alim background, he was educated in the religious sciences at an early age. While religious learning was often begun in a madrasa or masjid starting in the early teens, Ibn Abi Asim had a head start relative to his time period.

Eventually, Ibn Abi Asim left Basra for the city of Isfahan, further to the east. Late in life, he was granted a position as a qadhi at his new city of residence.

Ibn Abi Asim died in Isfahan in the year 900. He was 81 years old and at the time of his death, he was still holding his position as a qadhi. According to Iranian historian Abu Nu`aym, Ibn Abi Asim was buried in Isfahan's Doshabaz cemetery.

7. Al-Khatib al-Baghdadi

Abū Bakr Aḥmad ibn ʿAlī ibn Thābit ibn Aḥmad ibn Māhdī al-Shafi`ī,
(commonly known as al-Khaṭīb al-Baghdādī),
(Arabic: الخطيب البغدادي),
(10 May 1002 – 5 September 1071; 392 AH-463 AH),
was a historian.

Scientific Contributions

Al-Baghdādī wrote over 80 titles, of which are the following:

- Al-Kifaya fi ma'rifat usul 'ilm al-riwaya: an early work dealing with Hadith terminology, which Ibn Hajar praised as influential in the field.
- Al-Djami' li-akhlak al-rawi wa-adab al-sami.
- Al-Sabik wa 'l-lahik: dealing with hadith narrators of a particular type.
- Takyid al-'ilm: Questions whether putting traditions into writing is forbidden.
- Sharaf ashab al-hadith: Centers around the significance of traditionalists.
- Al-Mu'tanif fi takmilat al-Mu'talif wa 'l-mukhtalif: Correct spelling and pronunciation of names.
- Al-Asma' al-mubhama fi 'l-anba' al-muhkama: identifying unnamed individuals mentioned in hadith.
- Al-Rihla fi talab al-hadith.
- Al-Muttafik wa 'l-muftarik.
- Talkhis al-mutashabih fi 'l-rasm wa-himayat ma ashkala minhu min nawadir al-tashif wa 'l-wahm.
- Iktida' al-'ilm al-'amal.

- Ta'rīkh Madīnat al-Salām: or Ta'rīkh Baghdād wa Dhaīlih wa-l-Mustafād (تاريخ مدينة السلام) (تاريخ بغداد) وذيله والمستفاد) 'The History of Baghdād,' or Madīnat as-Salām ('City of Peace') and Appendix of Scholars - 23 volumes.

Biographers Sibt ibn al-Jawzi, Ibn Kathīr, and Ibn Taghribirdi wrote that the original was a work by as-Suri which al-Baghdādī had extended. Yāqūt al-Ḥamawī attributed the authorship to as-Surī's sister and accused al-Baghdādī of plagiarism, whereas Ibn Kathīr made no accusation of plagiarism, but attributed the original to as-Suri's wife. Abu'l-Faraj ibn al-Jawzi accused him of dishonesty in relation to the Ḥadīths.

Ibn Hajar declared his works influential in the field of the Science of hadith and Hadith terminology saying, "Scarce is the discipline from the disciplines of the science of ḥadīth on which he has not written a book." He then quoted Abu Bakr ibn Nuqtah, a Hanbali scholar, as saying, "Every objective person knows that the scholars of ḥadīth coming after al-Khaṭīb are indebted to his works."

Biographical Summary

Al-Khatib al-Baghdadi was born on 24 Jumadi' al-Thani, 392 A.H/May 10, 1002, in Hanikiya, a village south of Baghdad. He was the son of a preacher and he began studying at an early age with his father and other shaykhs. Over time he studied other sciences but his primary interest was hadith. At the age of 20 his father died and he went to Basra in search for hadith. In 1024 he set out on a second journey to Nishapur and he collected more ahadith in Rey, Amol and Isfahan. It is unclear how long he traveled but his own accounts have him back in Baghdad by 1028.

While he was an authority on hadith it was his preaching that led to his fame that would help him later in life. One biographer, Al-Dhahabi, said that contemporary teachers and preachers of tradition would usually submit what they had collected to Al-Baghdadi before they used them in their lectures or sermons.

Al-Khatib al-Baghdadi died on 5 September 1071 AD.

Al-Baghdadī originally belonged to the Hanbali school of Fiqh but later adopted Shafi'i school of fiqh. It is unclear if his change of allegiance followed a trip to Nahrawan in 1038, but in any case it provoked hostility from some Hanbalites. Despite the threat, under the protection of Caliph Al-Qa'im, al-Baghdadī lectured on ḥadīth in the Manṣūr Mosque.

When a rebellion in 1059 led by the Turkish general Basasiri deposed Caliph Al-Qa'im, and deprived Al-Baghdadi of his protection in Baghdad, he left for Damascus and there spent eight years as a lecturer at the Umayyad Mosque until a major controversy erupted. According to his biographers, Yaqut, Sibt ibn al-Jawzi, al-Dhahabi, as-Safadi, and Ibn Taghribirdi, this controversy involved al-Baghdadi's relationship with a youth, who, apparently had travelled with him from Baghdad. Yaqut relates that when news of the controversy reached the ruler of Damascus, he ordered that al-Baghdadi should be killed. However, the police chief, a Sunni, realizing that to follow the order would lead to a backlash against the Shi'i, warned al-Baghdadi to flee to the protection of Shari ibn Abi al-Hasan al-'Alawi. Al-Baghdadi spent about a year exiled in Sur, Lebanon, before he returned to Baghdad, where he died in September 1071. He was buried next to Bishr al-Hafi.

8. Abd al-Ghani al-Maqdisi

al-Imam al-Hafidh Abu Muhammad Abdul-Ghani ibn Abdul-Wahid bin Alī bin Surūr Ibn Rāfi' bin Hussain bin Ja'far al-Maqdisi al-Jammā'īlī al-Hanbali

(for short 'Abd al-Ghanī ibn 'Abd al-Wāḥid al-Jammā'īlī al-Maqdisi),

(Arabic: عبدالغني المقدسي),

(1146 – 1203 AD),

was a prominent mohaddith.

Scientific Contributions

The works of Abd al-Ghani al-Maqdisi include:

- Kitāb ut-Tawḥīd
- Akhbār Ad-Dajjāl
- Al-`Itiqād - A short text that outlines the foundational aqida.
- Al-Jāmi' as-Saghīr Li Ahkām al-Bashīr an-Nadhīr
- I`tiqād ul-Imām Ash-Shafi`ī -The author shows the complete agreement between all the Imams on foundational kalam and particularly the Imam's dislike for speculative theology.
- Al-Ahkām
- Al-Arba'īn Min Kalām Rabbil-Aalamīn
- Amr bi-l-Maʿrūf wa-n-Nahy ʿani-l-Munkar
- At-Targhīb fid-Du'ā al-Hathth Alayhi
- At-Tawakkul was Su'āl Allāh Azza wa Jall
- Al-Aathār al-Mardiyyah Fī Fadā'il Khayr il-Bariyyah
- Al-Iqtiṣād fil-I'tiqād-This is a book on advanced theology that itemises creed into a series of themes.
- Al-Miṣbaḥ fī `Uyun il-Aḥādith aṣ-Ṣiḥaḥ

- Mukhtaṣar Sīrah an-Nabī wa Sīrah Aṣḥabihi al-'Asharah (Short Biographies of the Prophet and His Ten Companions who were given the Tidings of Paradise)
- Ṭuḥfat ut-Ṭālibīn fīl Jīhad wal-Mujāhidīn
- Umdat ul-Aḥkām min Kalām Khayr il-Kalām
- Umdat ul-Aḥkām al-kubrā - Extended version of Umdat ul-Aḥkām min Kalām Khayr il-Kalām.
- Faḍā'il ul-Hajj
- Faḍā'il us-Ṣadaqah
- Faḍā'il Ashar Dhil-Hijjah
- Faḍā'il Umar bin al-Khattāb
- Faḍā'il Makkah
- Al-Kamāl Fī Ma'rifat ir-Rijāl, a collection of biographies of hadith narrators within the Islamic discipline of biographical evaluation of Ravis
- Miḥnah Imām Aḥmad bin Ḥanbal

Biographical Summary

Abd al-Ghani al-Maqdisi was born in 541 AH (1146 CE) in the village of Jummail in Palestine. He studied with scholars in Damascus, many of whom were from his own extended family.

He studied with the Imam of Tasawwuf, Shaykh Abdul Qadhir al-Jilani.

Abd al-Ghani al-Maqdisi was the first person to establish a school on Mount Qasioun near Damascus.

He died in 600 AH (1203 CE).

He was a relative of Diya al-Din al-Maqdisi, as his mother and Diya al-Din al-Maqdisi's grandmother were sisters. He had three sons named Muhammad, Abdullah and Abdur-Rahman, all of whom became prominent ulama. The alim, Ibn Qudamah al-Maqdisi was the maternal cousin of Abdul-Ghani, and Ibn Qudāmah described his association with Abdul-Ghani as: My friend in childhood and in seeking knowledge, and never did we race to goodness except that he would precede me to it, with the exception of a small number of occasions.

Abd al-Ghani al-Maqdisi was the author of Al-Kamal fi Asma' al-Rijal, a collection of biographies of hadith narrators within the Islamic discipline of biographical evaluation of Ravis.

9. Mohammed ibn Qasim al-Tamimi

Abu Abd Allah Mohammed ibn Qasim ibn Abd ar-Rahman ibn al-Karim al-Tamimi al-Fasi

(Arabic: محمد بن قاسم التميمي),

(born 1140/5, died 1207/8),

was a mohaddith and biographer from Morocco.

Scientific Contributions

Mohammed ibn Qasim al-Tamimi was the author of Al-Mustafad fi manaqib al-ubbad bi-madinat Fas wa ma yaliha min al-bilad. This book comprises 81 biographies of Moroccan sufis.

He also wrote a fahrasa in which he recorded the names of his teachers and the works he studied under them, called An-Najm al-mushiqa (The resplendent Star).

Biographical Summary

Al-Tamimi hailed from the Banu Tamim tribe which settled in al-Maghreb and al-Andalus.

He studied under Abu Madyan. There are also many references to At-Tamimi in the work of Muhyī al-Dīn Ibn al-Arabi.

Al-Tamimi was born in 1140 AD, and he died in 1207.

10. Al-Nawawi

Abū Zakariyyā Yaḥyā ibn Sharaf al-Nawawī

(Arabic: أبو زكريا يحيى بن شرف النووي),

(popularly known as al-Nawawī or Imam Nawawī),

(631–676 A.H./1234–1277),

was a Shafi'ite qadhi and mohaddath. He authored numerous and lengthy works ranging from hadith, to ilm-al-kalam, biography, and fiqh. Al-Nawawi never married.

Scientific Contributions

During his life of 44 years, he wrote "at least fifty books" including the following:

- Al Minhaj bi Sharh Sahih Muslim (شرح صحيح مسلم), making use of others before him, and is considered one of the best commentaries on Sahih Muslim. It is available online.
- Riyadh as-Saaliheen (رياض الصالحين), collection of hadith on ethics, manners, and conduct, popular in the Muslim world.
- al-Majmu' sharh al-Muhadhab (المجموع شرح المهذب), is a comprehensive manual of Islamic law according to the Shafi'I school. It has been edited with French translation by van den Bergh, 2 vols., Batavia (1882–1884), and also published at Cairo (1888).
- Minhaj al-Talibin (منهاج الطالبين وعمدة المفتين في فقه الإمام الشافعي), a classical manual on Shafi'i fiqh.
- Tahdhib al-Asma wa'l-Lughat (تهذيب الأسماء), edited as the Biographical Dictionary of Illustrious Men chiefly at the Beginning of Islam (Arabic); (F. Wüstenfeld (Göttingen, 1842–1847)).
- Taqrib al-Taisir (التقريب والتيسير لمعرفة سنن البشير النذير), an introduction to the study of hadith, it is an extension of Ibn al-

Salah's Muqaddimah; (was published at Cairo, 1890, with Suyuti's commentary "Tadrib al-Rawi"). (It has been in part translated into French by W. Marçais in the Journal asiatique, series ix., vols. 16–18 (1900–1901)).

- al-Hatt ala al-Mantiq (الحث على المنطق) - 'The Insistence upon Logic,' written to address epistemological and historical criticisms of logic.
- al-Arbaʿīn al-Nawawiyya (الأربعون النووية) - 'Forty Hadiths,' collection of forty-two fundamental traditions, frequently published along with numerous commentaries.
- Ma Tamas ilayhi hajat al-Qari li Saheeh al-Bukhaari (ما تمس إليه حاجة القاري لصحيح البخاري)
- Tahrir al-Tanbih (تحرير التنبيه)
- Kitab al-Adhkar (الأذكار المنتخبة من كلام سيد الأبرار); collection of supplications of prophet Muhammad.
- al-Tibyan fi adab Hamalat al-Quran (التبيان في آداب حملة القرآن)
- Adab al-fatwa wa al-Mufti wa al-Mustafti (آداب الفتوى والمفتي والمستفتي)
- al-Tarkhis fi al-Qiyam (الترخيص بالقيام لذوي الفضل والمزية من أهل الإسلام)
- Manasik (متن الإيضاح في المناسك) on Hajj rituals.
- Sharh Sunan Abu Dawood
- Sharh Sahih al-Bukhari
- Mukhtasar at-Tirmidhi
- Tabaqat ash-Shafi'iyah
- Rawdhat al-Talibeen
- Bustan al-`arifin

Some recent editions of the works of Imam Nawawi are given below.

Recent English language edition:

- Bustan al-ʿarifin (The Garden of Gnostics), Translated by Aisha Bewley

Editions of Minhaj al-Talibin:

- Minhaj et talibin: A Manual of Muhammadan Law ; According To The School of Shafi, Law Publishing Co (1977) ASIN B0006D2W9I
- Minhaj et talibin: A Manual of Muhammadan Law ; According To The School of Shafi, Navrang (1992) ISBN 81-7013-097-2
- Minhaj Et Talibin: A Manual of Muhammadan Law, Adam Publishers (2005) ISBN 81-7435-249-X

Editions of The Forty Hadith:

- Al-Nawawi Forty Hadiths and Commentary; Translated by Arabic Virtual Translation Center; (2010) ISBN 978-1-4563-6735-0
- Ibn-Daqiq's Commentary on the Nawawi Forty Hadiths; Translated by Arabic Virtual Translation Center; (2011) ISBN 1-4565-8325-5
- The Compendium of Knowledge and Wisdom; Translation of Jami' Uloom wal-Hikam by Ibn Rajab al-Hanbali translated by Abdassamad Clarke, Turath Publishing (2007) ISBN 0-9547380-2-0
- Al-Nawawi's Forty Hadith, Translated by Ezzeddin Ibrahim, Islamic Texts Society; New edition (1997) ISBN 0-946621-65-9
- The Forty Hadith of al-Imam al-Nawawi, Abul-Qasim Publishing House (1999) ISBN 9960-792-76-5
- The Complete Forty Hadith, Ta-Ha Publishers (2000) ISBN 1-84200-013-6
- The Arba'een 40 Ahadith of Imam Nawawi with Commentary, Darul Ishaat

- Commentary on the Forty Hadith of Al-Nawawi (3 Vols.), by Jamaal Al-Din M. Zarabozo, Al-Basheer (1999) ISBN 1-891540-04-1

Editions of Riyad al-Salihin:

- Gardens of the righteous: Riyadh as-Salihin of Imam Nawawi, Rowman and Littlefield (1975) ISBN 0-87471-650-0
- Riyad-us-Salihin: Garden of the Righteous, Dar Al-Kotob Al-Ilmiyah
- Riyadh-us-Saliheen (Vol. 1&2 in One Book) (Arabic-English) Dar Ahya Us-Sunnah Al Nabawiya.

Biographical Summary

Imam Nawawi was born in 1234 AD at Nawa near Damascus, Syria. The last part of his name refers to his hometown.

Yasin bin Yusuf Marakashi, says: "I saw Imam Nawawi at Nawa when he was a youth of ten years of age. Other boys of his age used to force him to play with them, but Imam Nawawi would always avoid the play and would remain busy with the recitation of the Noble Qur'an. When they tried to domineer and insisted on his joining their games, he bewailed and expressed his no concern over their action. On observing his sagacity and profundity, a special love and affection developed in my heart for young Nawawi. I approached his teacher and urged him to take exceptional care of this lad as he was to become a great religious scholar. His teacher asked whether I was a soothsayer or an astrologer. I told him I am neither soothsayer nor an astrologer but Allah caused me to utter these words. His teacher conveyed this incident to Imam's father and in keeping in view the learning quest of his son, decided to dedicate the life of his son for the service and promotion of the cause of Islam.

Imam Nawawi studied in Damascus from the age of 18 and after making the pilgrimage in 1253, he settled there as a private scholar.

During his stay at Damascus, he studied from more than twenty teachers who were regarded as masters and authority of their subject field and disciplines they taught. An-Nawawi studied Hadith, fiqh, its principles, syntax and Etymology. His teachers included Abu Ibrahim Ishaq bin Ahmad AI-Maghribi, Abu Muhammad Abdur-Rahman bin Ibrahim Al-Fazari, Radiyuddin Abu Ishaq Ibrahim bin Abu Hafs, Umar bin Mudar Al-Mudari, Abu Ishaq Ibrahim bin Isa Al-Muradi, Abul-Baqa Khalid bin Yusuf An-Nablusi, Abul-Abbas Ahmad bin Salim Al-Misri, Abu Abdullah Al-Jiyani, Abul-Fath Umar bin Bandar, Abu Muhammad At-Tanukhi, Sharafuddin Abdul-Aziz bin Muhammad Al-Ansari, Abul-Faraj Abdur-Rahman bin Muhammad bin Ahmad Al-Maqdisi, and Abul-Fada'il Sallar bin Al-Hasan Al Arbali among others.

He did ta'wil on some of the Qur'an verses and ahadith on the attributes of Allah. He states in his commentary of a hadith that:

This is one of the "hadiths of the attributes," about which scholars have two positions. The first is to have faith in it without discussing its meaning, while believing of Allah Most High that "there is nothing whatsoever like unto Him" (Qur'an 42:11), and that He is exalted above having any of the attributes of His creatures. The second is to figuratively explain it in a fitting way, scholars who hold this position adducing that the point of the hadith was to test the slave girl: Was she a monotheist, who affirmed that the Creator, the Disposer, the Doer, is Allah alone and that He is the one called upon when a person making supplication (du'a) faces the sky--just as those performing the prayer (salat) face the Kaaba, since the sky is the qibla of those who supplicate, as the Kaaba is the qibla of those who perform the prayer--or was she a worshipper of the idols which they placed in front of themselves? So, when she said, In the sky, it was plain that she was not an idol worshiper.

Nawawi drew the ire of Mamluk Sultan Rukn al-Din Baybars, when he petitioned on behalf of residents of Damascus who sought relief from heavy tax burdens during a drought that lasted many years. This prompted Baybars to threaten to expel him from Damascus. To this, he responded: "As for myself, threats do not harm me or mean anything to me. They will not keep me from advising the ruler, for I believe that this is obligatory upon me and others."

He died in 1277at Nawa at the relatively young age of 44.

Imam Nawawi's Forty Hadith were taught in the Mosque-Madrassa of Sultan Hassan in Cairo, Egypt.

An-Nawawi's lasting legacy is his contribution to hadith literature through his momentous works: Forty Hadiths and Riyadh as-Saaliheen. This made him respected in all madhabs, despite of him being of Shafi'i fiqh. According to Al-Dhahabi, Imam Nawawi's concentration and absorption in academic love gained proverbial fame. He had devoted all his time for learning and scholarship. Other than reading and writing, he spent his time contemplating on the interaction and complex issues and in finding their solutions.

In 2015, during the ongoing Syrian Civil War, his tomb was demolished by rebels linked to Al Nusra.

Fuqaha and Qadhis

Ideally, a mujtahid should specify **fiqh**, which is the set of rules for judicial decision making, and a set of such decisions. However, the five Imams who ventured into specifying the five fiqhs, were not Mujtahids. Therefore, these fiqhs are not ideal. Consequently, they differ among themselves sufficiently to warrant five different shades of Islam.

A **faqih** is one who can use the rules and decisions set of fiqh to make judicial decisions in a living court.

A **qadhi** is a faqih who is officially appointed to make binding judicial decisions that are enforced under the state authority.

In this volume, we use the terms Fiqh, faqih, and qadhi. Many translators have used jurisprudent, jurist, and judge in their place; however, the reader should be aware that they are not equivalent. We will therefore prefer to use the Islamic terminology in our discourse in English language.

The term **qiyasist** represents scientists who investigate judicial inferences and precedences. It is developed chiefly by the people of the Fiqh. They analyzed Quran and compiled ahadith, incorporating precedence, to decide judiciously. They use direct qiyas (analogies) with respect to the material in precedents. Qiyasists try not to use their intellectual interpolation and extrapolation in making a decision, so the personal discretion of the qadhi is not involved; therefore, most qadhis would likely be onboard in commending such a decision.

Qiyas is somewhat similar to the practice of judicial precedence in constitutional judicial systems. There is, however, an important difference, namely, only those occurrences of precedence are incorporated that happened in early Islamic era and which have been verified and validated by the factologists.

The term **istihsanist** is used and developed by the people of the Fiqh. Scripture, compiled ahadith, and precedents are used, as is the case with a qiyasist. Application of qiyas solves many of the judicial cases. However, in some judicial cases the application of qiyas does not lead to a unique decision, because in such cases the processes of qiyas do not converge. It can happen for many reasons. First, there may exist no precedence because the world conditions have changed in terms of the possible set of actions and events. An example would be the performance of rituals like prayer, fasting, and hajj during air travel. Second, the application of qiyas may not yield a clear and unique judgement because the set of qiyas processes available in a judicial case and related circumstances may be varied and diverse. In general, when the application of qiyas is insufficient to arrive at a unique judicial decision, the Qadhi needs to resort to the use of his own intellect to arrive at an istihsan with respect to what the correct judgement should be in a particular situation. In this process, the personal discretion of the qadhi is involved in such a way that benefit of doubt is amply allowed in favor of the accused. This is the essence of 'istihsan'.

In the case of qiyas the use of intellect is not specific to the particular qadhi since most qadhis would be onboard. However, that is generally not the case in the exercise of istihsan, so that the istihsan-based judgement may be more specific to the particular qadhi. Such distinction and finetuning points to the sophistication of Islamic judicial system. No such distinction and corresponding sophistication exists regarding the use of precedence among the European justice systems, and that can and does lead to arbitrariness and systematic bias.

11. Umm al-Darda

Umm al-Darda al-Kubra

(Arabic: أم الدرداء الكبرى),

was a companion of prophet Muhammad. She was a prominent faqih during the 7th century in Damascus.

Scientific Contributions

One of Umm al-Darda's students, ʿAbd al-Malik ibn Marwān, was the 5th Umayyad caliph. He studied fiqh under Umm al-Darda. The 14th-century Muslim historian Ibn Khaldun states, "ʿAbd al-Malik ibn Marwan is one of the greatest Caliphs. He followed in the footsteps of ʿUmar ibn al-Khattab, the Commander of the Believers, in regulating state affairs."

Umm Darda, taught in both Damascus, in the great Umayyad Mosque, and Jerusalem. Her class was attended by Imams, jurists, and Hadith ulama.

As an orphan child, she prayed with men, until she reached maturity at which point, she started praying with women. She became a teacher of hadith and fiqh and lectured in the men's section of the Mosque (something that was normally not allowed).

Being passionately devoted to teaching, Umm al-Darda has been teaching a large number of students. One day one student asked her about having many students: Have we wearied you? On that she answered:

> You (pl.) weary me? I have sought worship in everything. I did not find anything more relieving to me than sitting with ulama and exchanging knowledge with them.

Umm al-Darda had shown the piety, modesty and unpretentious-ness both in her daily life and teaching; asking no fee for delivering knowledge, and *living on the basis of charitable gifts.*

Umm Ad-Darda' was learned in the sciences of hadith. Imam Bukhari referred to her as an authority in sahih al Bukhari: "Umm Darda ... she was an expert mutakallim." Ibn Adbul Barr calls her "*an excellent alim among women, and a woman intellectual, being extremely religious and pious.*" (al isti'ab fi asma' al as hab).

Umm Ad-Darda' was held by Iyas ibn Mu`awiyah, himself an important alim of hadith of the time and a judge of undisputed ability and merit, to be superior to all the other hadith scholars of the period, including the celebrated masters of hadith like Al-Hasan Al-Basri and Ibn Sirin.

Umm Darda, taught in both Damascus, in the great Umayyad Mosque, and Jerusalem; as well as in her house. Her class was attended by Imams, qadhis, and Hadith ulama. Even Abdul Malik b. Marwah would attend the classes of Umm Darda, even though he had a teaching license from Abdullah bin Umar.

Ahmad ibn Hanbal related from Zayd ibn Aslam that he said:

> Abd al-Malik used to send an invitation to Umm Darda and she would spend as his guest, and he would ask her questions about the Prophet peace be upon him. He said, 'He arose one night and called his maid servant but she came slowly and he cursed her, so she (the maid) said, Do not curse, for indeed Abud-Darda related to me that he heard the Messenger of Allah peace be upon him say, "*Those who curse will not be witnesses or interceders on the Day of Judgement.*"

She issued a fatwa, which is still used today, allowing women to pray in the same sitting position (tashahhud) as men.

Biographical Summary

Umm al-Darda was a companion of prophet Muhammad. She was a prominent faqih and mohaddith during the 7th century in Damascus.

She was an orphan under the guardianship of Abul Darda. As a child, she used to sit with male scholars in the mosque, praying in men's rows and studying Quran with them.

She was the wife of the famous Mohaddith Abu Darda'.

She was regarded by Ibrahim ibn Abi Ablah as a pious and modest woman. In Ibn 'Asakir 's Ta'rikh madinat Dimashq, Tarajim al-nisa, is written that: I saw Umm al-Darda in Jerusalem sitting among poor women. A man came and distributed some money among them. He gave Umm al-Darda a fals (a copper). She said to her servant: Buy camel meat with it. Is not that money sadaqah? Umm al-Darda said: it came to us unasked.

Note of identity:

Some narrations describe two women, each going by the name Umm al-Darda. So, they name them as Umm al-Darda al-Kubra and Umm -alDarda al-Sughra. When the descriptions given for each of them is studied, they appear to be the same!

Mohammad Akram Nadwi in Al-Muhaddithat (the Women Scholars of Hadith) in the Index section, repeats the same references for the two Umm al-Darda's: so, he links the narrations of Umm Darda as-Sughra with Umm al-Darda al-Kubra. Thus, it could be stated, that:

> there was one woman named Umm al-Darda, an eminent hadith and fiqh scholar teaching in the mosques of Damascus and Jerusalim, lived in 7th century and joined the respectful attitude from the caliph 'Abd al-Malik ibn Marwan.

A center for teaching Quran, hifz and tajwid to women has been established in Bahrain in the name of Umm al-Darda.

12. Urwah ibn Zubayr

'Urwah ibn al-Zubayr ibn al-'Awwam al-Asadi

(Arabic: عروة بن الزبير بن العوام الأسدي),

(born 644, died 713),

was among the seven fuqaha who formulated the fiqh of Medina in the time of the Tabi'in.

Scientific Contributions

Urwah wrote many books, but destroyed them the day of the Battle of al-Harrah. He later had a feeling of regret, saying "I would rather have them in my possession than my family and property twice over." At the same time, he quashed any fears that they might become sources of authority alongside the Qur'an.

He is also known to have written one of the first writings in the area of the biography of the Prophet, known as the Tract of Seerah. This is not extant either but is known through Ibn Ishaq.

His narrations are transmitted by Ibn Shihab al-Zuhri.

Biographical Summary

He was the son of Zubayr ibn al-Awwam and Asmā' bint Abu Bakr. He was also the brother of Abd Allah ibn al-Zubayr and the nephew of Aisha bint Abu Bakr.

His most loved son was Hisham ibn Urwah.

He purportedly built a residential complex on some farming land on the outskirts of Medina, some 3km west of Masjid an-Nabawi.

He was born in the early years of the caliphate of Uthman in Medina and lived through the civil war which occurred after Uthman's martyrdom. Although his brother Abd-Allah ibn al-Zubayr wrested the rule from Abd al-Malik, it is unknown if he assisted him. He devoted himself to the study of fiqh and hadith and had the greatest knowledge of hadiths narrated from Aishah. He said, "Before Aishah died, I saw that I had become one of four authorities. I said, 'If she dies, there will be no hadith which will be lost from those she knows. I have memorized all of them."

13. Sa'id ibn Jubayr

Sa'id ibn Jubayr

also known as Abū Muhammad,

(Arabic: سعيد بن جبير),

(665–714),

was originally from Kufa, Iraq. He was regarded as one of the leading members of the Tabi'in. Sa'īd is held in the highest esteem by scholars of Hadith, and was one of the leading fuqaha of the time.

Scientific Contributions

Sa'id ibn Jubayr was a faqih and a Mohaddith.

Ibn Abi Mughira said that when people of Kufa visited Ibn `Abbas they used to ask him for Fatwa, he used to say to them: "Isn't Sa'id Ibn Jubayr among you?".

Ibn Hajar al-Asqalani, a 15th century writer states: He (Sa'id ibn Jubayr) narrated hadiths from

- Ibn Abbas, Ibn Al-Zubayr,
- Ibn Umar,
- Ibn Maqal,
- Udayy Ibn Hatim,
- Abu Mas`uod al-Ansari,
- Abu Sa`id al-Khudri,
- Abu Hurayra,
- Abu Musa al-Ash`ari,
- al-Dahhak ibn Qays al-Fihri,
- Anas,
- `Amr ibn Maymun,
- Abu `Abdulrahman al-Sulami

- `A'isha

`Amr ibn Maymun said that his father said that Sa'id ibn Jubayr passed away and everyone on the earth attained his knowledge.

Abu al-Qasim al-Tabari said: "He is a reliable Imam and hujjah on Muslims".

Ibn Hibban said: "He was faqih, worshiper, righteous and pious".

From him (Sa'id ibn Jubayr) are recorded Ahadith by Imams: Bukhari, Muslim, al-Tirmidhi, al-Nasa'i, Abu Dawud, Ibn Maja, Imam Ahmad ibn Hanbal, and Imam Malik ibn Anas.

Sa'id ibn Jubayr narrates 147 traditions in Sahih Bukhari and 78 in Sahih Muslim.

Biographical Summary

Sa'id ibn Jubayr was born in 665 AD. At the battle of Jamājim in 82 AH (699-701), Ibn al-Ash'ath and his followers, including 100,000 from amongst the mawāli, took on the army of al-Hajjāj (d. 714), the governor of the Iraqi provinces during the reign of the Umayyad caliph al-Walid I.

Within the forces of Ibn al-Ash'ath was a group known as the 'Battalion of Qur'an Reciters' headed by Kumayl ibn Ziyad an-Nakha`i. Sa`īd ibn Jubayr was among them.

The revolt was brutally put down and Sa`īd was forced to flee to the outskirts of Mecca. He persisted in travelling to Mecca itself twice a year to perform the hajj and `umrah and would enter Kufa secretly to help resolve peoples' religious issues.

Sa'id ibn Jubayr died in May 714 AD.

Sa'īd was finally apprehended and brought before al-Hajjāj. Excerpts from a transcript of their dialogue follows:

> Sa'īd ibn Jubayr entered upon al-Hajjāj, who asked his name (and he knew his name well):

Sa'īd: Sa'īd ibn Jubayr.

Al-Hajjaj: Nay, you are Shaqīy ibn Kusayr. (al-Hajjāj is playing with words here: Sa'īd means happy and Shaqī means unhappy; Jubayr means one who splints broken bones and Kusayr means one who breaks them.)

Sa'īd: My mother knew better when she named me.

Al-Hajjāj: You are wretched (shaqīta) and your mother is wretched" (shaqiyat). Then he told him: "By Allah, I will replace your dunya with a blazing Fire".

Sa'īd: If I knew you could do it, I would take you as a God.

Al-Hajjāj: I have gold and wealth. Bags of gold and silver were brought and spread before Sa'īd ibn Jubayr in order to try him.

Sa'īd: O Hajjāj, if you gathered it to be seen and heard in showing off, and to use it to avert others from the way of Allah, then by Allah, it will not avail you against Him in any way. Saying this, he aligned himself towards Qiblah.

Al-Hajjāj: Take him and turn him to other than the Qiblah. By Allah, O Sa'īd ibn Jubayr, I will kill you with a killing with which I have not killed any of the people.

Sa'īd: O Hajjāj choose for yourself whatever killing you want, by Allah you will not kill me with a killing except that Allah will kill you with a like of it, so choose for yourself whatever killing you like.

Al-Hajjāj: Turn him to other than the Qiblah.

Sa'īd: Wherever you [might] turn, there is the Face of Allah.

Al-Hajjāj: Put him under the earth.

Sa'īd: From it (the earth) We created you, and into it We will return you, and from it We will extract you another time.

Al-Hajjāj was outdone and ordered the beheading of Sa'īd ibn Jubayr.

Sa'īd was martyred in the month of Sha'bān, 95 AH (ca. May 714) at the age of 49.

Al-Hajjāj is reported to have soon lost his senses and died within a month.

14. Said ibn al-Musayyib

Abu Muhammad Sa'id ibn al-Musayyib ibn Hazn al-Makhzumi

(Arabic: سعيد بن المسيب),

(642–715),

was one of the foremost authorities of fiqh among the taba'een (generation succeeding the companions of Muhammad who are referred to as the sahaba). He was based in Medina.

Scientific Contributions

During the Battle of al-Harra and the subsequent takeover of Medina by the Syrian troops of the Umayyad caliph Yazid I in 683, Sa'id was the one Medinese who prayed in the Prophet's mosque. After Yazid died, he refused to take the oath of allegiance to the Mecca-based, anti-Umayyad caliph Abd Allah ibn al-Zubayr. After the Umayyad Abd al-Malik had reconquered the Caliphate, including Medina, he requested Sa'id marry his daughter (born of his marriage to Abu Hurayra's daughter) to Abd al-Malik's son and future caliph Hisham. Sa'id refused and, in the face of increasing pressures and threats, he offered her to Ibn Abi Wada', who stayed in the madrasa. In 705, Abd al-Malik commanded his governors to enforce the oath of allegiance to his son al-Walid I as his successor. Sa'id refused. Hisham ibn Isma'il al-Makhzumi, the governor of Medina, jailed him and had him beaten daily until the stick was broken, but he did not yield. When his friends, such as Masruq ibn al-Ajda' and Tawus, advised him to consent to al-Walid's caliphate to spare himself further torture, he answered: "People follow us in our actions. If we consent, how will we be able to explain this to them?" Hisham's successor Umar II (a maternal grandson of Umar), who governed Medina in 706-712, on the other hand consulted Sa'id in all of his executive decisions.

Those who received Islamic rulings and Traditions from Sa'id include Umar II, Qatadah, al-Zuhri and Yahya ibn Sa'id al-Ansari, and others.

Sa'id appears mainly to have argued from his own reasoning, by analogy, by the examples of Umar and Muhammad and by the Qur'an. He did not treat the hadith as a science with isnad (chains of transmission) in the way of those after him (especially al-Zuhri). As a result, many of his rulings have been equipped with spurious isnads and converted into hadiths.

It is similar with tafsir (Qur'anic interpretation): Sa'id argued his points from the Qur'an, but refused to expound on verses for their own context or meaning. To the extent a "tafsir of Ibn al-Musayyib" ever existed it was compiled by his students based on his rulings.

The leading fuqaha Malik ibn Anas and al-Shafi'I took as unquestionably authentic the hadiths that Sa'id narrated from Umar or Muhammad as authentic, without mentioning from whom he received them. In their view, Sa'id was of the same rank as the sahaba in knowledge and narration of hadiths.

Biographical Summary

Sa'id was born in 634, the son of al-Musayyib ibn Hazn of the Banu Makhzum clan of the Quraysh tribe. He was born during the caliphate of Umar (r. 634–644) and met most of the sahaba, including Umar's successors Uthman (r. 644–656) and Ali (r. 656–661).

Sa'id was well known for his piety, righteousness and profound devotion to Allah. He is renowned as the most eminent of The Seven Fuqaha of Medina. He began, as did Hasan al-Basri in Basra, to give opinions and deliver verdicts on legal matters when he was around twenty years of age. The Companions admired him greatly. On one occasion, Abdullah ibn Umar remarked, "If [Muhammad] had seen that young man, he would have been very pleased with him."

Sa'id married the daughter of Abu Hurayrah. He learned the hadiths that Abu Hurayrah narrated. Sa'id had his daughter play not with dolls, but with drums; later she learnt to cook.

Sa'id died in 715 AD.

15. Ibrahim al-Nakhai

Ibrahim ibn Yazid

(Arabic: إبراهيم بن يزيد;),

(better known as al-Nakhai (Arabic: النخعي)),

(c. 670–714),

was a prominent mutakammil and faqih. He was amongst the well-respected Tabi'un of Kufa.

Scientific Contributions

Al-Nakhai was an expert faqih and was amongst the leading ulama of Kufa. According to the 13th-century historian Ibn Khallikan (c. 1211–1282), al-Nakhai was also a 'celebrated doctor" of the city. He met many companions of Muhammad including, Anas ibn Malik and Aisha bint Abu Bakr. Al-Nakhai holds that Abu Bakr was the first male Muslim.

Biographical Summary

Ibrahim Al-Nakhai belonged to Kufa and was born c. 670. His father Yazid ibn al-Aswad was prominent member of the Nakha clan of Yemen. Ibrahim's mother was Mulayka bint Yazid ibn Qays, a sister of al-Aswad ibn Yazid. Ibrahim belonged to the Nakha, hence his laqab al-Nakhai. His eldest son was either named Ammar or Imran, thus his kunya Abu Ammar or Abu Imran.

Al-Nakhai died in 717 AD.

16. Raja ibn Haywah

Raja᾽ ibn Ḥaywa ibn Khanzal al-Kindī

(Arabic: رجاء بن حيوة),

(660 – after 724 AD),

was a prominent mutakallim and political adviser of the Umayyad caliphs Abd al-Malik (r. 685–705), al-Walid I (r. 705–715), Sulayman (r. 715–717) and Umar II (r. 717–720).

Scientific Contributions

He was a staunch defender of the religious conduct of the caliphs against their pious detractors. He played an important role in the construction of the Dome of the Rock in Jerusalem under Abd al-Malik. He became a mentor of Sulayman during the latter's governorship of Palestine and his secretary or chief scribe during his caliphate. He played an influential role in securing the succession of Umar II over Sulayman's brothers or sons and continued as a secretary to the new caliph. He spent the last decade of his life in retirement, though he maintained contact with Caliph Hisham (r. 724–743).

Raja played a key role in the construction of the Dome of the Rock in Jerusalem.

It was likely through the patronage of the Kindites in the caliphs' courts in Syria that Raja gained favor with the Umayyads, particularly Marwan's son and successor, Abd al-Malik (r. 685 –705. The latter entrusted Raja and his own Jerusalemite mawlā, Yazid ibn Sallam, with overseeing the financing of the Dome of the Rock's construction in Jerusalem. It is possible this was the reason for Raja's relocation to Palestine from the Jordan district and his new title sayyid ahl Filaṣtin (leader of the people of Palestine). Raja's role in its construction is described in the earliest known literary work specifically dedicated to the merits of Jerusalem, the Faḍāʿil al-Bayt al-Muqaddas written by the Jerusalemite preacher Ahmad al-Wasiti before 1019. Raja and Yazid were instructed by the caliph to spend generously on the building's construction and ornamentation. In an account recorded by the 15th-century Palestine-based historian Mujir ad-Din al-Ulaymi, Raja and Yazid informed Abd al-Malik that after the Dome of the Rock's completion there remained a surplus of 100,000 gold dinars in the construction budget. The caliph offered them the sum as an additional reward for their efforts, but both men refused; as a result, Abd al-Malik ordered that the coins be melted to gild the building's dome.

The historian Nasser Rabbat notes that several factors about Raja suggest he played a greater role in the founding of the Dome of the Rock, beyond fiscally managing its construction. On the one had he had a social connection to Palestine, attributed expertise about the holy sites of Jerusalem, and played an important role in developing the early Muslim tradition about Jerusalem's sanctity. On the other hand, he held a senior position in the Umayyad court and possessed knowledge of the Qur'an. Accordingly, Raja may have advised Abd al-Malik to choose the site of the Dome of the Rock on the Temple Mount and formulated the Qur'anic inscriptions which decorate the structure's interior and exterior.

Toward the end of the Dome of the Rock's completion in 691/92, Raja was assigned by Abd al-Malik to a joint embassy with the up-and-coming commander al-Hajjaj ibn Yusuf to negotiate a reconciliation with Zufar ibn al-Harith al-Kilabi, the Qarqisiya (Circesium) based rebel leader of the Qaysi tribes. The latter had given their allegiance to Abd al-Malik's anti-Umayyad rival, the Mecca-based caliph Abd Allah ibn al-Zubayr, and since their rout by the Umayyads and their Kindite and Banu Kalb allies at the Battle of Marj Rahit in 684, had launched a revolt throughout the al-

Jazira (Upper Mesopotamia) and the Syrian Desert. Raja displayed his moderate disposition by praying alongside Zufar when al-Hajjaj refused to do so. According to al-Baladhuri, Raja later interceded with Abd al-Malik to pardon the rebels who had participated in the mass anti-Umayyad, Iraqi rebellion of Ibn al-Ash'ath, a prominent Kufa-based Kindite, in 700–701.

When Abd al-Malik appointed his son Sulayman governor of Palestine, he assigned Raja as his mentor. Raja accompanied Abd al-Malik's son and successor al-Walid I (r. 705–715) on the Hajj (pilgrimage) to Mecca and Medina in 710.

By the time Sulayman acceded to the caliphate in 715, Raja had gained a reputation as the ascetic of the Umayyads and the "outstanding man of religion of his age for Syria". He related traditions from certain companions of the Islamic prophet Muhammad, including Mu'awiya, Jabir ibn Abd Allah, Abu Umama al-Bahili and Abd Allah ibn Umar, which were, in turn, related by numerous later Muslim traditionists. In a quote attributed to Sulayman's brother Maslama, the head Umayyad commander on the Byzantine front, "through Raja and his likes, we are rendered victorious". In a testament to Raja's loyalty to the Umayyad caliphs Sa'id ibn Jubayr (d. 714) stated, Raja "used to be regarded as the most knowledgeable faqih in Syria, but if you provoke him, you will find him Syrian in his views quoting Abd al-Malik ibn Marwan saying such-and-such."

Raja served as Sulayman's chief kātib (secretary or scribe) and head of the administration of justice. He is credited by the Mamluk historian Ibn Fadlallah al-Umari for advising Sulayman, while he was governor of Palestine, to select the site of Ramla as the new capital of Islamic Palestine, replacing nearby Lydda (Lod). Raja played an influential role in securing the succession of Sulayman's paternal cousin, the son of Abd al-Aziz ibn Marwan, Umar II, to the caliphate over expectations in the Umayyad ruling family that one of Sulayman's brothers or sons would accede. In the account of the historian al-Waqidi (d. 823), while Sulayman was on his deathbed at his army camp in Dabiq during the major offensive against the Byzantines in 717, Sulayman's succession became a pressing issue. Abd al-Malik had formally designated al-Walid and Sulayman as his successors, but did not specify anyone beyond them; nonetheless, his intention that the office of the caliphate remain in the hands of his direct descendants was common knowledge in the ruling family. Sulayman's chosen successor, his eldest son Ayyub, had predeceased him and the ill caliph debated potential replacements with Raja.

The two Umayyad factions present at Dabiq were an anonymous group of Sulayman's inner circle represented by Raja and the family of Abd al-Malik, apparently represented by the caliph's brother Hisham. The latter faction favored another of Sulayman's brothers, Yazid II, who was away on the Hajj pilgrimage, to succeed, while the former favored Umar. In al-Waqidi's accounts, which are ultimately traced back to Raja's own account of the events, Raja persuaded Sulayman to bypass his own sons and brothers in favor of Umar. Raja was chosen to execute Sulayman's will. He secured the decision by securing oaths of allegiance from the Umayyad family to Sulayman's willed successor whose name was kept secret in a sealed letter. Once he gained their oaths, Umar was revealed as the next caliph and Yazid II as the next in line. He threatened the use of force against Sulayman's brothers following their protestations at being bypassed.

Raja first met Umar during the Hajj pilgrimage of 710, when Umar served as governor of Medina for al-Walid. During Umar's caliphate (717–720), Raja was one of the caliph's three kātibs.

Biographical Summary

Raja, known also by his kunya "Abū'l-Miqdām" or "Abū Naṣr", was the son of a certain Haywa ibn Khanzal. He was born in Beisan in the Jordan district before moving south to Palestine.

According to a report traced to Raja and recorded by the Egyptian historian al-Suyuti (d. 1505), Raja ultimately considered himself a Jerusalemite. His approximate year of birth was c. 660, during the early reign of the first Umayyad caliph, Mu'awiya I (r. 661–680).

The 9th-century historian Khalifa ibn Khayyat mentions that Raja was a mawlā (non-Arab, Muslim client or freedman) of the Kinda. There exists his full genealogy which places him as a great-grandson of the Kindite tribesman Imru' al-Qays ibn Abis, a contemporary of the first caliph Abu Bakr. Moreover, Imru al-Qays and Raja both lived in Baysan, the former in his later life and Raja in his youth. Because of his family's residence in the Palestine or Jordan district of Syria, Raja is occasionally given the nisba (epithet) of al-Filasṭīnī ("the Palestinian") or al-Urdunnī ("the Jordanian"). The family likely hailed from or settled in an area inhabited by their Kindite tribal kin or patrons, whose prominence in Syria had grown under Mu'awiya and further still under Caliph Marwan I (r. 684–685).

Following the death of Umar, Raja likely entered retirement. According to the Persian historian Abu Nu'aym al-Isfahani (d. 1038), he refused to accompany Umar's successor, Caliph Yazid II (r. 720–724) on the latter's visit to Jerusalem. Caliph Hisham (r. 724–743) wrote to Raja expressing regret about his executions of the Qadari (at the time a theological school of Islam that asserted humans possessed free will) scholars Ghaylan al-Dimashqi and Salih Qubba. To that Raja wrote back supporting Hisham's decision; the executed scholars had been known political dissidents during the reign of Raja's patron, Umar II. According to the historian Ibn al-Athir (d. 1233), Raja died in Qussin, a place in Kufa's environs.

17. Abu Yusuf

Yaqub ibn Ibrahim al-Ansari

(Arabic: يعقوب بن إبراهيم الأنصاري),

(better known as Abu Yusuf (Arabic: أبو يوسف)),

(729 – 798 AD),

was a student of jurist Abu Hanifah (d.767) who helped spread the influence of the Hanafi school of fiqh through his writings and the government positions he held.

Scientific Contributions

Abu Yusuf served as the qadhi al-qudhat during reign of Harun al-Rashid. His most famous work was Kitab al-Kharaj, a treatise on taxation and fiscal problems of the state.

During his lifetime, Abu Yusuf created a number of literary works on a range of subjects including fiqh, international law, narrations of collected traditions (ahadith), and others. The Kitāb al-Fihrist, a bibliographic compilation of books written by various authors in the 10th century, by Ibn al-Nadim, mentions numerous titles authored by Abu Yusuf. With one exception, none of these works listed in the Fihrist have survived.

The exception is his book entitled Kitāb al-Kharāj, a treatise on taxation and financial issues facing the empire written at the request of the caliph, Harun al-Rashid. The Islamic empire was at the height of its power at the time of his writing and in his treatise, he sought to advise the caliph on how to appropriately conduct financial policies in accordance with religious law. While the caliph took some suggestions and ignored some, the overall effect was to limit the ruler's discretion over the tax system.

A selection of other works credited to him that do not appear in the Fihrist have also survived. The Kitab al-Athar is a collection of Kufian traditions (ahadith) which he narrated. Kitab Ikhtilaf Abi Hanifa wa Ibn Abi Layla is a comparison of the opinions between the legal authorities, Abu Hanifa and Abu Layla. Kitab al-Radd 'Ala Siyar al-Awza'i is a "reasoned refutation with broad systematic developments," of the opinions regarding the laws of war of the famous Syrian scholar, al-Awza'i. Some excerpts from his various other works that have not survived in their totality were incorporated in texts written by his disciples and were passed on through succeeding generations. For example, excerpts from Abu Yusuf's book, Kitabal-Hiyal (Book of Legal Devices) were incorporated in the book, Kitabal-Makharidj fi 'l-Hiyal written by his disciple, Muhammad al-Shaybani.

As a disciple of Abu Hanifa, Abu Yusuf's doctrine largely presupposes that of his mentor. His writings and prominent political positions helped advance the Hanafi school of fiqh throughout the Islamic empire. While most of his legal opinions (fatwas) were firmly rooted in the doctrine and methodology espoused by his former teacher, there are some points on which he diverged and revealed his own legal thought. The doctrine of Abu Yusuf was more dependent on traditions (ahadith) than his master, in part because there were more authoritative prophetic traditions available to him in his time.

He also reacted against the somewhat unrestrained reasoning exhibited by Abu Hanifa. However, he was not always consistent; in a certain number of cases, he disregarded sounder and more highly developed doctrine by diverging from the opinions of his former teacher. Based on his surviving works and opinions, certain tendencies in Abu Yusuf's reasoning have been determined, such as his tendency to logically follow the implications of a proposition to an absurd conclusion, and his use of rather caustic language in his attacks on opponents' positions, and in defense of his own.

Abu Yusuf is also noted for the frequency in which he changed positions on various issues, which has been suggested is a result of his experience as a qadhi.

Abu Yusuf's greatest legacy is in affirming and advancing the Hanafi fiqh as the predominant source of legal thought in the Islamic empire and providing a legal framework for defining and restricting caliphal authority in regard to fiscal policy.

Following is a partial list of the works of Abu Yusuf'.

- Kitab al-Kharaj, his most famous work, is a treatise on taxation and fiscal problems of the state prepared for the caliph.
- Usul al-fiqh - the earliest known work of principles of fiqh. A portion of his works were devoted to international law.
- Kitab ul-Aathar, a collection of traditions (ahadith) he narrated.
- Kitab Ikhtilaf Abi Hanifa wa Ibn Abi Layla, one of the early works on comparative fiqh.
- Kitab al-Radd ‘Ala Siyar al-Awza’i, a refutation of the famous Syrian faqih and Mohaddith, al-Awza’i, on the law of war.

Some of these books were published by Al Ihya Al Ma'arif an N'omaniya under the guidance of Abul Wafa Al Afghani.

Biographical Summary

Abu Yusuf lived in Kufa and Baghdad, in what is now Iraq, during the 8th century. His genealogy has been traced back to Sa'd b. Habta, a youth in Medina in the time of the Prophet, and his birth date is estimated based on the date of his death to be around 729 CE.

Abu Yusuf was raised poor but with a ferocious appetite for knowledge. His mother disapproved of his academic desires, insisting that he master some trade (the art of tailoring, according to some source) so as to help make ends meet. He complied with his mother's wishes, but also kept up his academic studies. His talent and commitment was eventually recognized by Abu Hanifa who became his mentor with Abu Yusuf as his star pupil. He is portrayed as an incredibly studious individual who was intent in his pursuit for knowledge and legal understanding.

It has been verified that he studied religious law and traditions in Kufa and Medina under a number of scholars including Abu Hanifa, Malik b. Anas, al-Layth b. Sa'd and others. Under the guidance of Abu Hanifa, Abu Yusuf achieved incredible success and helped develop and spread the influence of the Hanafi school of Islamic law through his writings and the government positions he held.

Abu Yusuf lived in Kufa until he was appointed Qadhi in Baghdad. According to one narration, Abu Yusuf was able to provide sound advice pertaining to religious law to a government official who rewarded him generously and recommended him to the caliph, Harun al-Rashid. What is known is that Abu Yusuf became a close acquaintance of Abbasid caliph, Harun al-Rashid, who eventually granted him the title of Grand Qadhi, or Qadhi 'l-qudhat; the first time such a title had been conferred upon someone in Islamic history.

The Caliph frequently consulted Abu Yusuf on legal matters and financial policy and even bestowed upon him the ability to appoint other Qadhis in the empire. Abu Yusuf held the position of Grand Qadhi until his death in 798CE.

18. Ibn al-Qasim

'Abd ar-Rahman ibn al-Qasim al-'Utaqi,

(also known as Ibn al-Qasim),

(750–806 AD),

was a prominent early faqih in the Maliki school from Egypt. He was one of Malik's main companions and had a tremendous influence in recording the positions of the school.

Scientific Contributions

Ibn al-Qasim kept the company of Malik for about twenty years; and it was from Malik that he learned his fiqh. Ibn al-Qasim transmission of the Muwatta is considered to be the soundest transmission.

When Malik was asked about Ibn al-Qasim and Ibn Wahb, he replied that Ibn Wahb was a knowledgeable man whilst Ibn al-Qasim was a true faqih).

Sahnun wrote Mudawwana, the most comprehensive collection of Maliki fiqh; and Sahnun learned its content from Ibn al-Qasim.

Ibn al-Qasim has the same position in the Maliki school as Muhammad al-Shaybani has in the Hanafi school, in so far as both of them transmitted their respective schools and made free use of ijtihad (independent reasoning).

Biographical Summary

Ibn al-Qasim was born in 750 AD in Egypt in a mosque known as the 'Utaqi Mosque, at a time when the Abbasids took control of the Muslim world from the Umayyads. Ibn al-Qasim's origins were from the Palestinian town of Ramla. He was a descendant from the slaves of Ta'if whom the Prophet Muhammad had freed. Ibn al-Qasim's father was in the Dewan, and he used the money he inherited from him for his studies.

He travelled from Egypt to Medina after what is recorded as a visionary dream and after having been drawn to gatherings of religious knowledge in Egypt. In Medina, he met Malik as well as Ibn Wahb, another of Malik's famous companions. Ibn al-Qasim kept the company of Malik for about twenty years; and it was from him that he learned his fiqh. In Medina he also met Al-Layth, Ibn al-Majishun and Muslim ibn Khalid al-Zanji. Many people narrated from him, and consulted him about Malik's fatwas. Ibn Wahb used to say, "If you want this business – meaning the fiqh of Malik – you must have Ibn al-Qasim. He is unique in it".

He was known as having ascetic qualities and spent much of his time reciting the Quran such that he would finish many readings in a short space of time. On his return to Egypt, he refused to marry the daughters of wealthy officials and generally kept clear of the ruling class. He died in Egypt at the age of 63 in the month of Safar, 806 CE, three days after returning from a trip to Mecca. Ibn al-Qasim left behind him two sons Abd ar-Rahman and 'Umar.

19. Mā Shā' Allāh ibn Athari

Mā Shā' Allāh ibn Athari

(Arabic: ما شاء الله إبن أثري),

(c. 740–815 AD),

was a pious jurisprudent from Khorasan.

Scientific Contributions

Mā Shā' Allāh ibn Athari was a pious jurisprudent. That is according to The Kitāb al-Fihrist by Ibn Al-Nadim (c.998) (Arabic: كتاب الفهرست) (The Book of Catalogue) which is an encyclopedia of all scholars and their scholarly works. The works attributed to Mā Shā' Allāh ibn Athari by Ibn al-Nadim are as follows:

- (كتاب المواليد الكبير) The Big Book of Births (14vols);
- (الواحد والعشرين في قرانات والأديان والملل) The Twenty-One On Conjunctions, Religions and Sects;
- (مطرح الشعاع) The Projection of Rays;
- (المعاني) The Meaning;
- (صنعة الإسطرلابات والعمل بها) Construction and Operation of Astrolabes;
- (ذات الحلق) The Sphere of Truth;
- (الأمطار والرياح) Rains and Winds;
- (السهمين) The Two Arrows;
- (الحروف) The Letters;
- (السلطان) The Sultan;
- Book known as The Seventh & Decimal
  - Ch.1 - (ابتداء الأعمال) The Beginning of Actions;
  - Ch.2 - (على دفع التدبير) Averting What Is Predestined;
  - Ch.3 - (في المسائل) On Issues;

   - Ch.4 - (شهادات الكواكب) Testimonies of the Stars;
   - Ch.5 - (الحدوث) Happenings;
   - Ch.6 - (تسيير النيرين وما يدلان عليه) Movement and Indications of the Two Luminaries [sun & moon]);
- (السفر) The Journey;
- (الأسعار) Perceptions;
- (المواليد) Nativities;
- (تحويل سني المواليد) Transfer of the Years of Nativities;
- (الدول و الملل) Governments (Dynasties) and Sects;
- (الحكم على الاجتماعات والاستقبالات) Judgement on Gatherings and Receptions;
- (المرضى) The Sick;
- (الصور والحكم ليها) Constellations (Ṣūr) and its Judgements.

Unlike the claims, (such as by Thomas Hockey et al. (eds.). *The Biographical Encyclopedia of Astronomers*, *Springer*, 2007, pp. 740-741) the above publications are not on Astrology or Astronomy except few. We will here go with the original sources like Ibn-al-Nadim, and consider the Recent Westerners' descriptions as repudiated fabrications, motivated by their agenda.

Biographical Summary

Originally from Khorasan he lived in Basra. The bibliographer al-Nadim in his Fihrist, described him as *a leader in jurisprudence and virtuous*. The original sources of his time are reliably unambiguous on this.

Many Western writers, however, seem to have turned him into somebody he was not, for reasons arguably based on their agenda. For example, Western writers interpret jurisprudence as the science of judgments of the stars, and call him a Jewish Astronomer who founded

Cosmology in Arabia, and was also the leader of Astronomy and Cosmology at Darul Hikmah, as well as a court astrologer (Wikipedia).

As there is no support for such fabrications in the original sources, we here have adopted the original sources.

20. Sahnun

Sahnun ibn Sa'id ibn Habib at-Tanukhi

(Arabic: سحنون بن سعيد بن حبيب التنوخي),

(c. 776/77 – 854/55) (160 AH – 240 AH),

was a jurist in the Maliki school from Qayrawan in modern-day Tunisia.

Scientific Contributions

Sahnun's greatest contribution to Muslim scholarship was Al-Mudawwana, a compendium of the legal opinions of the school of Medina as stated by Imam Malik, after the death of the Imam. The compilation and revision process involved four imams of the Maliki school: Asad ibn al Furat (d.213 AH); Al-Ashhab (d.204); Ibn al-Qasim (d.191 AH); and Sahnun himself. It is referred to as "al Umm" of the Maliki school.

Sahnun's revision and transmission of the Mudawwana was the major factor in the spread of the Maliki school across the Islamic West.

Sahnun was known for his strong orthodoxy, even to the point of refusing to pray behind a Mu'tazilite imam. He excluded heretical sects from the mosque, including the Ibadi, Mu'tazilites and others. The Encyclopedia of Islam states:

> Hitherto, in the multiple circles of ulama, representatives of all tendencies were able to express themselves freely in the Great Mosque of Kairouan. In a process amounting to a purging of the community of scholars there, Sahnun put an end to this "scandal". He dispersed the sects of the ahl al-bida; the leaders of heretical sects were paraded ignominiously, and some were compelled to recant in public. Sahnun was one of

the architects of the exclusive supremacy of Sunnism in its Maliki form throughout the Muslim West.

Biographical Summary

His original name was Abdu Salaam Ibn Said Ibn Habib (عبد السلام بن سعيد بن حبيب). He gained the nickname 'Sahnun' (a type of sharp bird) because of his quickness of mind.

His father was a soldier from Homs in Syria; and was from the Arab tribe of Tanukh.

In his youth Sahnun studied under the scholars of Qayrawan and Tunis. In particular, he learned from the Tripolitanian scholar `Ali bin Ziyad, who had learned from Imam Malik. Sahnun wanted to reach Malik, but could not afford the financial means to travel, and Malik died before Sahnun could travel. In 178 AH he traveled to Egypt to study under students of Malik. Later on, he continued to Medina and studied under other prominent scholars, returning to North Africa in 191 AH.

At the age of 74 Sahnun was appointed qadhi of North Africa by the Aghlabid emir Muhammad I Abul-Abbas. He had refused the appointment for a year, only accepting after the emir swore to give him a free hand in matters of justice, even if this involved prosecuting members of the emir's family and members of his court. Upon accepting the appointment, he was said to have told his daughter Khadija, "Today your father has been slain without a knife."

Sahnun held the emir to this promise. He was scrupulous in his judgments. He was courteous towards litigants and witnesses, but he was strict towards the men surrounding the emir. For example, he refused to allow the aristocrats to send representatives on their behalf in litigations;

and he even refused a direct request from the emir not to interfere in the ventures of the aristocrats.

Sahnun died in Rajab of 256 AH (854 AD). The men surrounding the emir refused to join his funeral prayer, due to his harshness against them. Yet the emir conducted the funeral prayers in person. People of Qayrawan were greatly aggrieved by his passing.

21. Yahya ibn Aktham

Abu Muhammad Yahya ibn Aktham

(Arabic: أبو محمد يحيى بن أكثم),

(died 857),

was a ninth century faqih. He twice served as the chief qadhi of the Abbasid Caliphate, from ca. 825 to 833 and 851 to 854.

Scientific Contributions

Yahya ibn Aktham is generally characterized as having been affiliated with the Hanafis, and many state this without specifying his teacher in Fiqh. However, some do state that he learnt from Waki' ibn al-Jarrah in particular, who would give juridical opinions on the position of Abu Hanifah, and that he also related Hadith reports from one of Abu Hanifa's main students, Muhammad al-Shaybani. Ibn Hazm's view is that he was part of an independent Basran Ra'y tradition that was later subsumed by the Hanafi school. Al-Daraqutni further alternatively lists him as a Shafi'i but this is doubted by primary sources.

Yahya enjoyed strong relations with the caliph and became an immensely influential member of the administration. All decisions made by the viziers were reportedly first submitted to him for approval.

In 831 he participated in al-Ma'mun's campaign against the Byzantines and was put in command of a raiding party which set out from Tyana. The following year he accompanied the caliph to Egypt and briefly acted as qadhi there.

In 851, following the abandonment of Mu'tazilism by al-Mutawakkil (r. 847–861), Yahya was again made chief qadhi and he moved to Samarra. During his judgeship he appointed a mix of qadhis, selecting both men who had formerly been affiliated with Mu'tazilism, as well as those who appealed to the orthodox Hanbalis. He remained chief qadhi until July 854.

Following are some of the modern references to the works of Yahya ibn Aktham.

- Bosworth, C.E. *(2002).* "Yahya b. Aktham". *The Encyclopedia of Islam, New Edition, Volume XI: W–Z. Leiden and New York: BRILL. p. 246.* ISBN 90-04-12756-9.
- *Hinds, M. (1993).* "Mihna". *The Encyclopedia of Islam, New Edition, Volume VII: Mif–Naz. Leiden and New York: BRILL. pp. 2–6.* ISBN 90-04-09419-9.
- Ibn Khallikan, Shams al-Din Abu al-'Abbas Ahmad ibn Muhammad *(1871).* Ibn Khallikan's Biographical Dictionary, Vol. IV. *Trans. Baron Mac Guckin de Slane. Paris: Oriental Translation Fund of Great Britain and Ireland.*
- Al-Kindi, Muhammad ibn Yusuf *(1912). Guest, Rhuvon (ed.).* The Governors and Judges of Egypt *(in Arabic). Leyden and London: E. J. Brill.*
- Al-Mas'udi, Ali ibn al-Husain *(1873).* Les Prairies D'Or, Tome Septieme. *Ed. and Trans. Charles Barbier de Meynard and* Abel Pavet de Courteille. *Paris: Imprimerie Nationale.*
- Melchert, Christopher *(1997).* The Formation of the Sunni Schools of Law, 9th-10th Centuries C.E. *Leiden: Brill.* ISBN 90-04-10952-8.
- Melchert, Christopher *(1996). "Religious Policies of the Caliphs from al-Mutawakkil to al-Muqtadir: AH 232-295/AD 847-908". Islamic Law and Society. 3 (3): 316–342.* doi:10.1163/156851996259906 9. JSTOR 3399413.

- *Stewart, Devon J. (2004).* "Muhammad b. Jarir al-Tabari's Al-Bayan 'an Usual al-Ahkam and the Genre of Usul al-Fiqh in Ninth Century Baghdad". *In Mont-gomery, James E. (ed.). 'Abbasid Studies: Occasional Papers of the School of 'Abbasid Studies, Cambridge, 6-10 July 2002. Leuven: Peeters.* ISBN 90-429-1433-5.
- Al-Tabari, Abu Ja'far Muhammad ibn Jarir *(1985–2007). Ehsan Yar-Shater (ed.). The History of Al-Ṭabarī. Vol. 40 vols. Albany, NY: State University of New York Press.*
- Tillier, Mathieu. (2009). Les cadis d'Iraq et l'État abbasside (132/750-334/945). Damascus: Institut français du Proche - Orient, 2009.

Biographical Summary

Yahya was born in Marw in Khurasan and was a member of the Banu Tamim; he himself claimed to be a descent from the qadhi Aktham ibn Sayfi. He studied hadith and fiqh in Basra. In 817-8 he was appointed as qadhi (judge) of Basra, and he held that position until 825.

Following his dismissal from Basra, Yahya was selected by al-Ma'mun (r. 813–833) to serve as chief justice (qadhi al-qudat).

By the end of al-Ma'mun's reign, however, Yahya had fallen out of favor, and he decided to return to Iraq. Throughout his career he had been forced to defend himself against consistent allegations of pederasty, and by the time of al-Ma'mun's death he was also facing accusations of financial mismanagement.

As a supporter of Sunni orthodoxy, he was also opposed to the Mu'tazilite belief that the Qur'an had been created, which put him at odds with the caliph's adherence to Mu'tazilism. Following the accession of al-

Ma'mun's brother al-Mu'tasim (r. 833–842), Yahya lost his position and was replaced with the Mu'tazilite Ahmad ibn Abi Du'ad.

In 854 AD al-Mutawakkil dismissed him in favor of Ja'far ibn Abd al-Wahid ibn Ja'far al-Hashimi. His money and land were also seized at the time of his dismissal, and he was placed under house arrest.

In 857 Yahya decided to go on the *pilgrimage and intended to take up residence in Mecca.* Upon learning that al-Mutawakkil had forgiven him, he changed his mind and set out to return to Iraq, but he died, during the journey, on April 857; he was buried in al-Rabadhah.

22. Abd Allah al-Qaysi

Abu Muhammad Abd Allah bin Muhammad bin Qasim bin Hilal bin Yazid bin 'Imran al-'Absi al-Qaysi

(Arabic: عبدالله القيسي),

(died 825 AD),

was a 9th century faqih and mutakallim.

Scientific Contributions

Abd Allah al-Qaysi left the Malikite fiqh for the Zahirite branch. He was listed by later Zahirite jurist Ibn Hazm as having been, along with Ruwaym, Ibn al-Mughallis, and Mundhir bin Sa'īd al-Ballūṭī, one of the primary proponents of the Zahirite school of Islamic law. Ibn Hazm, who was also an early champion of the school, was essentially reviving Ibn Qasim's efforts; earlier Zahirites such as Balluti kept their views to themselves.

Biographical Summary

Born in Islamic Spain, Ibn Qasim moved to Iraq for a time, and studied under Dawud al-Zahiri. Ibn Qasim died in the year 272 AH, corresponding to 885 or 886 on the Gregorian calendar.

23. Ahmad ibn Abu Bakr al-Zuhri

Abū Muṣʿab Aḥmad ibn Abī Bakr al-Qāsim ibn al-Ḥārith al-Zuhri (Arabic: أبو مصعب أحمد بن أبي بكر القاسم بن الحارث الزهري), (767–856 CE / 150–242 AH), was a faqih and qadhi. He was a student of Malik ibn Anas.

Scientific Contributions

He was born and lived in Medina, where he wrote a work called al-Mukhtaṣar fī al-fiqh ('The Epitome on Fiqh').

Al-Zuhri also wrote a recension of Malik ibn Anas' Kitāb al-Muwaṭṭaʾ. Al-Zuhri 's recension is five to ten percent larger than the recension of Yahya ibn Yahya al-Laythi, which is considered the standard version in the Maliki school of law.

In his fiqh opinions (fatwas), he relied not only on hadith reports, but also on rational discretion (raʾy). He was dismissed from his position as qadhi by Qutham ibn Ja'far in 210 AH (825/826 CE).

Biographical Summary

Al-Zuhri was born in 767 AD, and he died in 856 AD.

24. Al-Tahawi

Abu Ja'far Ahmad ibn Muhammad ibn Salamah ibn 'Abd al-Malik ibn Salamah, al-Azdi al-Hajari al-Misri al-Tahawi al-Hanafi

(for short Abu Ja'far Ahmad al-Tahawi),

(Arabic: أبو جعفر الطحاوي),

(or simply aṭ-Ṭaḥāwī (Arabic: الطحاوي)),

(843 – 5 November 933),

was a Hanafi faqih, mohaddith, and mutakallim from Egypt. He is known for his work al-'Aqidah al-Tahawiyyah, a summary of Sunni Islamic creed which influenced Hanafis in Egypt.

Scientific Contributions

Aṭ-Ṭaḥāwī was famed for his expertise in both ḥadīth and Ḥanafī fiqh even during his own lifetime, and many of his works, such as Kitāb Maʿāni al-Āthār and ʿAqīdah aṭ-Ṭaḥāwīyyah, continue to be held in high regard by Sunni Muslims today.

Many of aṭ-Ṭaḥāwī's contemporaries praised him and noted him as both a reliable mutakallim and mohaddith. He was widely held as a distinguished and prolific writer and became known as the most learned faqīh amongst the Ḥanafīs in Egypt, in addition to having knowledge of all the madhāhib. Over fifteen commentaries have been produced on his creedal treatise, ʿAqīdah aṭ-Ṭaḥāwīyyah, including shuruh by the Hanafi jurist Ismail ibn Ibrahim al-Shaybani and the Taymiyyan-inclined Ibn Abi al-Izz.

Aṭ-Ṭaḥāwī gained a vast knowledge of Hadīth in addition to Ḥanafī fiqh and his study circles consequently attracted many students of knowledge who related ḥadīth from him and transmitted his works. Among them were al-Da'udi, the head of the Zahiris in Khurasan, and aṭ-Ṭabarānī, well known for his biographical dictionaries of ḥadīth transmitters.

Aṭ-Ṭaḥāwī authored many other works, close to forty different books, some of which are still available today, including:

- Maʿāni al-Āthār (معاني الآثار)
- al-ʿAqīdah aṭ-Ṭaḥāwīyyah (العقيدة الطحاوية)
- Aḥkām al-Qur'ān al-Karīm (أحكام القرآن الكريم)
- Al-Mukhtaṣar fil-Furūʿ (المختصر في الفروع)
- Sharḥ Mushkil al-Āthār (شرح مشكل الآثار)
- Sharḥ Maʿāni al-Āthār (شرح معاني الآثار)
- Sharḥ al-Jāmiʿ al-Kabīr (شرح الجامع الكبير)
- Sharḥ al-Jāmiʿ aṣ-Ṣaghīr (شرح الجامع الصغير)
- Ash-Shurūṭ aṣ-Ṣaghīr (الشروط الصغير)
- Ash-Shurūṭ al-Kabīr (الشروط الكبير)
- Ikhtilāf al-ʿUlamā' (إختلاف العلماء)
- ʿUqūd al-Marjān fi Manāqib Abī Ḥanīfa an-Nuʿmān (عقود المرجان في مناقب أبي حنيفة النعمان)
- Tārīkh al-Kabīr (تاريخ الكبير)
- Ḥukm Arāḍi Makkah al-Mukarramah (حكم أراضي مكة المكرمة)

Biographical Summary

Aṭ-Ṭaḥāwī was born in the village of Ṭaḥā in upper Egypt in 229 AH (843 CE) to an affluent Arab family of Azdī origins. He began his studies with his maternal uncle, Ismāʿīl ibn Yaḥyā al-Muzanī, a leading disciple of ash-Shāfiʿī, but in 249 AH (863 CE), at approximately 20 years of age, aṭ-Ṭaḥāwī abandoned the Shāfiʿī school of fiqh in favour of the Ḥanafī school. Different versions are given by his biographers of his change to the Ḥanafī school, but the most probable reason seems to be that the system of Abū Ḥanīfa appealed to his critical insight more than that of ash-Shāfiʿī.

Aṭ-Ṭaḥāwī then studied under the head of the Ḥanafīs in Egypt, Aḥmad ibn Abī ʿImrān al-Ḥanafī, who had himself studied under the two primary students of Abū Ḥanīfa, Abū Yūsuf and Muḥammad ash-Shaybānī. Aṭ-Ṭaḥāwī then travelled to Syria in 268 AH (882 CE) for further studies in Ḥanafī jurisprudence and became a student of Abū Khāzim ʿAbd al-Ḥamīd ibn ʿAbd al-ʿAzīz, the chief qāḍi of Damascus.

He died on the 14th day of Dhū-l Qaʿdah, 321 AH (November 5, 933 CE), and was buried in al-Qarāfah, Cairo.

25. Al-Baqillani

Abū Bakr Muḥammad ibn aṭ-Ṭayyib al-Bāqillānī

(Arabic: أبو بكر محمد بن الطيب الباقلاني),

(for short often known as al-Bāqillānī , or reverentially as Imām al-Bāqillānī by Ash'arites),

(c. 950 - 5 June 1013),

was a famous mutakallim, faqih, and logician who spent much of his life defending and strengthening the Ash'ari school of thought in Islam.

Scientific Contributions

An accomplished rhetorical stylist and orator, al-Baqillani was held in high regard by his contemporaries for his expertise in debating ilm-al-kalam and fiqh issues. Al-Baqillani is often given the honorary epithets Shaikh as-Sunnah ("Doctor of the Prophetic Way"), Lisān al-Ummah ("Voice of the Community"), Imād ad-Dīn ("Pillar of the Faith"), Nāsir al-Islām ("Guardian of Islam"), and Saif as-Sunnah ("Sword of the Prophetic Way") by Ash'aris.

After acquiring expertise in both ilm-al-kalam and Maliki fiqh he expounded the teachings of the Ash'ari school, and taught Maliki fiqh in Baghdad. He held the office of chief Qadhi in Baghdad and in 'Ukbara, a town not far from the capital.

Al-Baqillani became a popular lecturer, and took part in debates with well-known scholars of the day. Because of his debate skills, the Amir 'Adud ad-Dawlah dispatched him as an envoy to the Byzantine court in Constantinople and he debated Christian scholars in the presence of their king in 371 AH/981 AD.

Al-Baqillani supported the doctrine of the apologetic miracle being proof of prophecy, the non-creation of the Qur'an, inter-cession, and the possibility of seeing God.

Ibn Taimiyyah called al-Baqillani 'the best of the Ash'ari mutakallimeen, unrivalled by any predecessor or successor'.

Fifty-five titles of works written by al-Baqillani have been listed, the great majority on fiqh and kalam matters, and many written against his opponents.

- Al-Inṣāf fīmā Yajib I'tiqāduh
- I'jāz al-Qur'ān (The Inimitability of the Qur'an)
- Al-Intiṣār lil-Qur'ān
- Al-Taqrīb wal-Irshād aṣ-Ṣaghīr
- Kitāb Tamhīd al-Awā'il wa-Talkhīṣ ad-Dalā'il
- Manāqib al-A'immah al-Arba'ah

Biographical Summary

Born in Basra in 330 AH / 950 CE, Al-Baqillani spent most of his life in Baghdad, and studied theology under two disciples of al-Ash'ari, Ibn Mujahid at-Ta'i and Abul-Hasan al-Bahili. He also studied jurisprudence under the Maliki scholars Abu 'Abdillah ash-Shirazi and Ibn Abi Zayd al-Qayrawani.

Al-Baqillani died in 403 AH / 1013 CE

26. Diya al-Din al-Maqdisi

Ḍiyā' al-Dīn Abu 'Abdallah Muhammad ibn 'Abd al-Wahid al-Sa'di al-Maqdisi al-Hanbali

(Arabic: ضياء الدين المقدسي),

(1173-1245 AD),

was a Hanbali faqih.

Scientific Contributions

Following are some works by Ḍiyā' al-Dīn al-Maqdisi.

- Al-Āhādith al-Jiyād al-Mukhtārah min mā laysa fī Ṣaḥīḥain: a collection of hadith arranged by the name of the Companion narrating each hadith, in alphabetical order. He was unable to complete it. He intended to include only authentic hadith a goal which, to a large extent, he accomplished.
- A short treatise, Ikhtisās al-Qur'ān Bi 'Awdihī ilā al-Rahīm al-Rahmān, a book bringing together the ahādīth and narrations pertaining to the Qur'an being erased from this Earth and returning to Allāh.
- As-Sunan wal-Ahkam `un il-Mustafa Alaihi Afdal us-Salati was-Salam
- Fada'il Al A'amaal: a collection of hadith highlighting the virtues of various actions, such as prayer, fasting, charity, and visiting the sick. His book is not to be confused with the similarly titled Fazail-e-Amaal by Muhammad Zakariyya al-Kandhlawi.

Biographical Summary

Diya' al-Din was born in Damascus in 1173. His parents had emigrated from Nablus in the crusader territory of Jerusalem shortly before his birth, along with 155 of other Hanbalis of the area, in response to threats against their shaykhs from the crusader lord of Nablus, Baldwin of Ibelin. Al-Dhahabi described him as the Sheikh of hadith scholars. He recorded Maqdisi's death in the year 1245 C.E.

He was a relative of Abd al-Ghani al-Maqdisi, as his grand-mother and Abd al-Ghani al-Maqdisi's mother were sisters. Ibn Qudamah was his maternal uncle.

27. Sidi Mahrez

Abu Mohamed Mahrez ben Khalaf ben Zayn

(Arabic: سيدي محرز بن خلف),

(951–1022 AD),

was a Tunisian Wali (Saint), scholar of the Maliki fiqh, and a Qadhi.

Scientific Contributions

He was a teacher of Maliki fiqh. He proposed to his teacher Ibn Abi Zayd al-Qayrawani (922 – 996 CE) to write a Aqidah and fiqh education book, and his proposal resulted in a book titled Risala fiqhiya.

Biographical Summary

He was born in 951 AD in Ariana to a father of Arab origin who traced his lineage to Abu Bakr. He studied in Kairouan and then in Fatimid-Egypt and became a teacher of Maliki fiqh upon his return. At the age of 57, he left his home-town (Ariana) and went into seclusion in Carthage. In c. 1014 he settled in Tunis, in a house in Bab Souika, which would become, after his death in 1022 AD, his mausoleum and later the Sidi Mahrez Mosque.

28. Al-Lakhmi

Ali ibn Muhammad Al-Qayrawani Al-Rab'i Al-Lakhmi
(also known as Imam al-Lakhmi),
(Arabic: علي بن محمد القيرواني الربعي اللخمي),
(c. 1006 – 1085 CE) (390 AH – 478 AH),
was a famous faqih in the Maliki fiqh.

Scientific Contributions

Al-Lakhmi's juristic compendium titled al-Tabsirah is an important text in the Maliki legal school. It is a commentary on one of the Maliki school's most famous work, al-Mudawwana, by Sahnun b. Sa'id (d. 240/854). In the book he discusses many questions in detail and develops opinions, relating these to teachings of other Maliki scholars.

Biographical Summary

His nisba indicates that he is from the Arab tribes of Banu Lakhm. He was one of the most important figures in the school and his opinions are still well known and respected to this day. Al-Lakhmi was one of four jurists whose positions were held as authoritative by Khalil ibn Ishaq in his book titled Mukhtassar (one of the most important of the later texts in the relied upon positions of the school).

Al-Lakhmi was born in 1006 AD in Qayrawan and spent the early part of his life there before moving to Sfax. Here he continued his education and then began teaching students in the mosque of the city. One of his students was al-Mazari (d. 536/1141), who refers to al-Lakhmi frequently. Al-Lakhmi died in Sfax in 478/1085.

29. Abu Bakr ibn al-Arabi

Abū Bakr Muḥammad ibn ʿAbdallāh ibn al-ʿArabī al-Maʿāfirī al-Ishbīlī (Arabic: أبو بكر محمّد ابن عبدالله ابن العربى المعافرى الأسفلى), (born in Sevilla in 1076 and died in Fez in 1148), was a qadhi and scholar of Maliki fiqh.

Scientific Contributions

Ibn al-'Arabi wrote many books on several different subjects, including hadith, fiqh, usul, Qur'an studies, adab, grammar and history. His works include "Book on the Arrangement of the Travel that Raised my Interests in Religions and Experiences of the Great Authorities and Eminent People by the Observer of Islam and the Various Lands".

Ibn al-'Arabi also wrote "The Rule of Interpretation", and "Protective Guards Against Strong Objections", a source of comments that al-Ghazali made to his students. Two of al-'Arabi's books, "Tartib al-rohla li al-targhib fi al-millah" and "Qanun al-ta'wil", provided descriptions of the al-Arabi's travels, and specifically recounted religious life in the holy city of Jerusalem. These accounts are important, as they may be the only eyewitness accounts by a Muslim in Jerusalem during the Seljuq period, and they also provide a critical Muslim perspective.

Ibn al-'Arabi continued to study, reflect upon, and challenge the works of al-Ghazali. For example, al-Ghazali believed that, "there is not in the sphere of possibility anything more excellent, more perfect or more complete than what God has in fact created." However, Ibn al-'Arabi argues that there is a limitation of God's right. We can see this argument by Ibn al-'Arabi in some of his other works. For example, there were (and probably still are) times when judges and lawyers were/are faced with a situation where there is no legal text or scripture to help provide insight or guidance on the judicial decision. In these cases, judges and lawyers must use their best discretion to determine the rule of law.

Laws of slander came into question, and defining the punish-ment as a right of God or a private right were debated. While Ibn al-'Arabi recognized that there are two views on whether the right is of God or a private right; ultimately, he felt that the crime should largely be seen as a private right, as it is conditioned by the victim filing a petition.

Ibn al-'Arabi reflected upon the nature of the soul and the study and theory of knowledge. Ibn al-'Arabi studied the Sufi argument that knowledge can only be achieved through purity of the soul, chastening of the heart, and an overall unity between the body and the heart, as well as removal from material motives. Ibn al-'Arabi argues that this is an extreme position, and believes rather that *there is no connection between knowledge a person acquires and any sacred or devout acts that his soul has performed.*

Ibn al-'Arabi used his knowledge of the soul in his studies of law and ethics. For example, when discussing abortion, madhhabs judgments differ considerably. Malikis and Hanafis tend to take opposite positions on this issue. *Malikis generally forbid induced miscarriage after conception, as this is seen to be the point at which the soul is breathed into the unborn child. While Hanafis hold that "induced miscarriage is not punishable until the 120th day of conception".* Ibn al-'Arabi tried to bridge the gap between the Maliki and the Hanafi opinions by "granting greater protection rights to the embryo after ensoulment," although ultimately, he did not succeed in bridging this gap.

Ibn al-'Arabi wrote on many other subjects. For instance, he wrote on the mistreatment and disciplining of women. He once wrote, "The [slaves] need to be disciplined with a stick, while the [free man] will not need more than an indication. Among women and even men, there are those who will behave well only through correction (adab). Any man who knows it has to resort to discipline [his wife], although it is preferable if he abstains from it." However, it seems that Ibn al-'Arabi was more focused on trying to express "beating in a non-violent way." He believed that this is the "only way allowed by the divine revelation," because the objective of beating in a non-violent way was ultimately to improve the wife's behavior.

Although Abu Bakr ibn al-'Arabi may have some critics, he was generally a highly acclaimed authority on hadith, and was regarded as being trustworthy and reliable.

His other Major books are:

- Commentary on Tirmidhi's Hadith Collection famously known as "'Aridhat al-Ahwazi'".
- Commentary on the Quran known as '"Ahkam al-Quran"'. It contains commentary on the legal rulings of the Qur'an according to the Maliki school.
- Al-'Awasim min al-Qawasim (العواصم من القواصم) or "Defense Against Disaster", is a history book that became famous for his strong reply against the Shia.

Biographical Summary

Abu Bakr Ibn al-'Arabi (born 468/1076, died 543/1148) was an "Andalusian Malikite qadhi". He was born in Seville Al-Andalus. Ibn al-'Arabi's father (Abu Muhammand ibn al-'Arabi) was a high ranking statesman working for the Taifa king of Seville, al-Mu'tamid ibn 'Abbad (r. 1069–1091). However, in 1091 when Al-Andalus was taken over by the Almoravids, Ibn al-'Arabi (now 16), and his father decided to leave for a less turbulent setting, (his father also had political motivations). The two al-'Arabis travelled by ship to Egypt, and from there they turned to Jerusalem, where they stayed from 1093 to 1096.

Al-'Arabi devoted himself to his studies, teaching, and writing.

After leaving Jerusalem in 1096, both al'-Arabis' traveled to Damascus and Baghdad to study. They settled in Baghdad and returned there after they performed pilgrimage. While in Jerusalem, Ibn al-'Arabi was enticed by all of the ulama he met there, and performing the hajj became an addition in his quest for knowledge. It was only when he returned to Baghdad in 1097 that Ibn al-'Arabi finally met Imam Abū Ḥāmed al-Ghazālī, under whom Ibn al-'Arabi studied, beginning at the age of 21.

Under al-Ghazali, an Islamic theologian, philosopher and Sufi mystic, Ibn al-'Arabi studied closely. As a result, Ibn al-'Arabi is said to be one of the "most important sources of information about al-Ghazali's life and his teachings". When it came to al-Ghazali's ilm-al-kalam, Ibn al-'Arabi became a master, and was enthusiastic, but perhaps more importantly critical of his teachings. Although Ibn al-'Arabi undoubtedly respected al-Ghazali, he was not afraid to express his feelings of difference when it came to the teachings of falsafa (philosophy).

After his father died in 1099 (at age 57), Ibn al-'Arabi, age 26, headed back to Seville. After being gone for 10 years studying in the Muslim east, he was an esteemed and credited scholar and teacher, as well as a main source to spread the works and teachings of al-Ghazali in the Muslim west.

Ibn al-'Arabi died in 1148 AD.

30. Qadhi Ayyad

Abu al-Fadl 'Iyad ibn Amr ibn Musa ibn 'Iyad ibn Muhammad ibn 'Abdillah ibn Musa ibn 'Iyad al-Yahsubi al-Sabti

(Arabic: أبو الفضل عياض بن موسى بن عياض بن عمرو بن موسى بن عياض بن محمد بن عبد الله بن موسى بن عياض اليحصبي السبتي),

(1083–1149, born in Ceuta),

was Maliki faqih and great imam of that city and, later, a qadhi in the Emirate of Granada.

Scientific Contributions

He was one of the most famous scholars of Maliki fiqh and author of the well-known Ash-Shifa, on the virtues of the prophet, and Tartib al-mardarik wa-taqrib al-masalik li-marifat alam madhab Malik, a collection of biographies of eminent Malikis.

Qadhi `Iyad's other well-known works include:

- Ikmal al-mu`lim bi fawa'id Muslim, a famous comment-ary on Sahih Muslim which transmitted and expanded upon al-Maziri's own commentary, al-Mu`lim bi-fawa'id Muslim. Qadhi `Iyad's own commentary was utilized and expounded upon heavily by Al-Nawawi in his own commentary of Sahih Muslim.
- Bughya al-ra'i lima Tadmanahu Hadith Umm Zara` min al-Fawa'id, published with Tafsir nafs al-Hadith by Al-Suyuti.
- al-I`lam bi Hudud Qawa'id al-Islam, written on the five pillars of Islam.
- al-Ilma` ila Ma`rifa Usul al-Riwaya wa Taqyid al-Sama`, a detailed work on the science of Hadith.

- Mashariq al-Anwar `ala Sahih al-Athar, based on al-Muwatta of Malik ibn Anas, Sahih Al-Bukhari of Imam Bukhari and Sahih Muslim by Muslim ibn al-Hajjaj.
- al-Tanbihat al-Mustanbata `ala al-kutub al-Mudawwana wa al-Mukhtalata.
- Daqa`iq al-akhbar fi dhikr al-janna wa-l-nar, a "eschatological manual" describing the joys of Jannah (heaven) and the horrors of jahannam (hell)

Biographical Summary

Iyaḍ was born into an established family of Arab origin in Ceuta. Son of a notable scholarly family, ʿIyad was able to learn from the best teachers Ceuta had to offer. The qadhi Abu ʿAbd Allah Muhammad b. ʿIsa (d. 1111) was ʿIyad's first important teacher and is credited with his basic academic formation. Growing up, ʿIyad benefited from the traffic of scholars from al-Andalus, the Maghrib, and the eastern Islamic world. He became a prestigious scholar in his own right and won the support of the highest levels of society.

In his quest for knowledge, Iyad spent part of 1113 and 1114 visiting Cordoba, Murcia, Almeria, and Granada. He received ijāzas from the most important traditionist of his time, Abū ʿAlī al-Ṣadafī (d. 1120) in Murcia, and met with some of the most celebrated scholars of the moment, such as Ibn al-Hajj (d. 1134), Ibn Rushd (d. 1126), and Ibn Hamdin (d. 1114).

ʿIyad was appointed qadhi of Ceuta in 1121 and served in the position until 1136. During his tenure as qadhi of Ceuta he was extremely prolific. Iyad's overall fame as a faqih and as a writer of fiqh (positive law) was based on the work he did in this city.

Iyad was also appointed the qadhi of Grenada where he worked for just over a year.

In doctrine, Iyad influenced later scholars like Ibn Taymiyyah and Taqī ad-Dīn as-Subkī (d.1355) in expanding the definition of heresy in apostasy. He is the first to call for the death penalty for those Muslims guilty of "disseminating improprieties about Muḥammad, or question-ing his authority in all questions of faith and profane life".

He headed a revolt against the coming of the Almohades to Ceuta, but lost and was banished to Tadla and later Marrakech. He was a student of Abu Abdillah ibn Isa, Abu Abdillah ibn Hamdin and Abu al-Hassan ibn Siraj.

He was a teacher of Averroes and Ibn Maḍāʾ.

He died in 1149. Sources disagree on how and where he died. Some sources, including one written by his son, Muhammad, describe how he ingratiated himself with the Almohads in Marakech and eventually died of sickness during a military campaign. Other sources describe how he died a natural death while acting as a rural qadhi near Tadla. Later sources tend to assume a violent death at the hands of the Almohads. Although he was opposed to the Almohads and the ideas of Ibn Hazm, he did not hold enmity for the Zahirite school of Sunni Islam, which the Almohads and Ibn Hazm followed. Ayyad's comments on Ibn Hazm's teacher Abu al-Khiyar al-Zahiri were positive, as was Ayyad's characterization of his own father, a Zahirite mutakallim.

Cadi Ayyad University, also known as the University of Marrakech, was named after him. Qadhi Ayyad is also well known as one of the seven saints of Marrakech and is buried near Bab Aïlen.

31. Al-Qurtubi

Abu 'Abdullah Muhammad ibn Ahmad ibn Abu Bakr al-Ansari al-Qurtubi

(Arabic: أبو عبدالله القرطبي),

(1233 – 1286 AD),

was an Andalusian faqih and muhaddith. He was taught by prominent scholars of Córdoba, Spain, and he is well known for his commentary of the Quran named Tafsir al-Qurtubi.

Scientific Contributions

Al-Qurtubi was very skilled in commentary, narrative, recitation and fiqh; as is clearly evident in his writings. The depth of his scholarship has been recognized by many scholars. In his works, Al-Qurtubi defended the Ash'ari point of view and criticized the Mu'tazilah.

The hadith scholar Dhahabi said of him, "..he was an imam versed in numerous branches of scholarship, an ocean of learning whose works testify to the wealth of his knowledge, the breadth of his intelligence and his superior learning."

Following are some works by Al-Qurtubi.

- Tafsīr al-Qurṭubī: the most important and famous of his works, this 20 volume commentary has raised great interest, and has had many editions. It is often referred to as al-Jamī' li-'Aḥkām. The commentary is a general inter-pretation of the whole of Quran with a Maliki point of view. Any claims made about a verse are stated and thoroughly investigated.
- al-Tadhkirah fī Aḥwāl al-Mawtà wa-Umūr al-Ākhirah (Reminder of the Conditions of the Dead and the Matters of the Hereafter): a book dealing with the topics of death, the punishments of the grave, the end times and the day of resurrection.

- Al-Asnà fi Sharḥ al-Asmā' al-Ḥusnà
- Kitāb ut-Tadhkār fi Afḍal il-Adhkār
- Kitab Sharḥ it-Taqaṣṣi
- Kitab Qam' il-Ḥirṣ biz-Zuhd wal-Qanā'ah
- At-Takrāb li-Kitāb it-Tamhīd
- al-Mufhim lima Ushkila min Talkhis Sahih Muslim

Biographical Summary

Al-Qurtubi was born in Córdoba, Al-Andalus in 1233 AD. His father was a farmer and died during a Spanish attack in 1230. During his youth, he contributed to his family by carrying clay for use in potteries. He finished his education in Cordoba, studying from renowned scholars ibn Ebu Hucce and Abdurrahman ibn Ahmet Al-Ashari. After Cordoba's capture in 1236 by king Ferdinand III of Castile, Al-Qurtubi left for Alexandria, where he studied hadith and tafsir. He then passed to Cairo and settled in Munya Abi'l-Khusavb where he spent the rest of his life. He is known for his modesty and humble lifestyle.

Al-Qurtubi died in 1286 AD, and was buried in Munya Abi'l-Khusavb, Egypt.

His grave was carried to a mosque where a mausoleum was built under his name in 1971, open for visiting today.

32. Ibn Daqiq al-'Id

Ibn Daqiq al-'Id

(Arabic: ابن دقيق العيد),

(1228–1302),

was an alim in the fundamentals of fiqh and aqeeda. He was born in Yanbu into the Arab tribe of Banu Qushayr,

Scientific Contributions

Ibn Daqiq al-'Id was an authority in the Shafi'i legal school. Although Ibn Daqiq al-'Id studied Shafi'i fiqh under Ibn 'Abd al-Salam, he was also proficient in Maliki fiqh.

Ibn Daqiq al-'Id served as chief qadhi of the Shafi'i school in Egypt.

Ibn Daqiq al-'Id, as a teacher of hadith, produced scholars of the next generation, like al-Dhahabi, al-Nuwayri, and other.

In his lifetime, Ibn-Daqiq wrote many books but his commentary on the Nawawi Forty Hadiths has become his most popular. In it he comments on the forty hadiths compiled by Yahya Al-Nawawi: al-Nawawi's Forty Hadith. His commentary has become so popular that it is virtually impossible for any scholar to write a serious book about the forty hadiths without quoting Ibn-Daqiq.

Biographical Summary

Ibn Daqiq al-'Id was born in 1228 AD, and he died in 1302 AD.

33. Taqi al-Din al-Subki

Abu Al-Hasan Taqī al-Dīn Ali ibn Abd al-Kafi ibn Ali al-Khazraji al-Ansari al-Subkī

(Arabic: أبو الحسن تقي الدين علي بن عبد الكافي بن علي الخزرجي الأنصاري السبكي),

(1284 – 1355 AD),

was a Shafi'i faqih, mohaddith, Qur'anic mufassir, and chief qadhi of Damascus.

Scientific Contributions

Having left Egypt in his youth, al-Subkī settled down in Syria where he rose through the ranks to the position of chief qadhi of Syria. He was the preacher of the Umayyad mosque at Damascus and a professor in several colleges.

He presided as chief qadhi for seventeen years, at the end of which he became ill, was replaced by his son Taj al-Din al-Subkī and returned to Cairo where he died in 1355 AD.

Subkī belonged to the Sunni Ash'ari school of theology and in line with his school strongly opposed anthropomorphism. He also vehemently defended the Ashari view that Paradise and Hell Fire are eternal and to that end wrote a comprehensive treatise entitled "Al-I'tibar" in which he stated that: "The doctrine of the Muslims is that the Garden and the Fire will not pass away. Abu Muhammad ibn Hazm has transmitted that this is held by consensus and that whoever opposes it is an unbeliever by consensus". Subkī reiterates this elsewhere in the treatise although he is careful to clarify that he does not label any particular person an unbeliever.

- Shifa' as-Siqam fi Ziarat khayr al'Anam (شفاء السقام في زيارة خير الأنام) - 'Cure for the Sick in Visiting the Best of Mankind' archive.org (in Arabic)
- Al-Sayf al-Saqil fi al-Radd ala Ibn Zafil (السيف الصقيل في الرد على ابن زفيل) - Refutation of Ibn al-Qayyim
- Al-Durra al-Mudiyya fi al-Radd 'ala Ibn Taymiyya (الدرة المضية في الرد على ابن تيمية) - Refutation to Ibn Taymiyya
- al-'Itibār bī baqā' al-janat wa'l-nār fi ar-rad 'ala ibn Taymiyah wa ibn al-Qiyam al-Qayilin bī fana' an-Nār. (الاعتبار ببقاء الجنة والنار في الرد على ابن تيمية وابن القيم القائلين بفناء النار) - Contemplation of the eternity of Paradise and Hell, A response to Ibn Taymiyah and Ibn al-Qayyim on the temporality of Hell.
- Naqid al-'Ijtimā' wa'l-'Iftirāq fī Masā'il al'Aymān wa't-Talāq (نقد الاجتماع والافتراق في مسائل الأيمان والطلاق) - 'Critique of Communion and Separation in Matters of Faith and Divorce.'
- Al-'Ashbāh wa'n-Naẓā'r (الأشباه والنظائر) - 'Analogues and Pairs' (in Arabic, 3 vols)
- 'Ibraz al-Hukam min hadīth rafa' al-Qalam (إبراز الحكم من حديث رفع القلم) - 'Illustration of ruling in hadith "Raising the Pen"'.

Biological Summary

Taqī al-Dīn al-Subkī was born in 1284 AD in the village of Subk in Egypt. He received his education in Cairo by such scholars as Ibn Rif'a in Sacred Law, al-Iraqi in Qur'anic tafsir and al-Dimyati in hadith. He also traveled to acquire knowledge of hadith from the scholars of Syria, Alexandria and the Hijaz. Eventually he taught at the Mansuriyya school located in the Ibn Tulun's mosque.

He returned to Cairo where he died in 1355 AD

34. Taj al-Din al-Subki

Abū Naṣr Tāj al-Dīn 'Abd al-Wahhāb ibn ʿAlī ibn 'Abd al-Kafi al-Subkī (Arabic: تاج الدين عبد الوهاب بن علي بن عبد الكافي السبكي),
(or Tāj al-Dīn al-Subkī (تاج الدين السبكي), or simply Ibn al-Subki),
(1327 – 1379 AD),
was a leading faqīh, a muḥaddith and a historian from the celebrated al-Subkī family of Shāfiʿī ʿulamā, during the Mamluk era.

Scientific Contributions

In his late twenties he began to assist his father as qāḍī (Chief judge) of Syria, and on his father's retirement to Cairo in 1354, he replaced him as qāḍī of Damascus. He also held the title of Mufti.

Following are some of the works of Tāj al-Dīn al-Subkī.

- Ṭabaqāt al-Šāfi'iyyaẗ - Kubrā, Wusṭā wa Ṣughrā (Large, Medium and Concise); Biographical dictionary of the scholars of the Shāfi'ī legal school; based on the Tabyīn kadhib al-Muftarī fī mā nusiba ilā al-Imām Abī al-Ḥasan al- Ash'arī of Ibn 'Asākir; (Cairo: Maṭba'aẗ al-Ḥusayn-iyyaẗ al-Miṣriyyaẗ, 1906).
- Kitāb Mu'īd an-Ni'am wa-Mubīd an-Niqām ("The restorer of favors and the restrainer of chastisements"); Arabic text with introduction and notes by David Vilhelm Myhrman: treats 113 trades, professions and offices of the author's own time, in the light of how their exponents should behave in order to recover God's favor. (English translation: Luzac & Co., London, 1908).
- Kitāb al-Ashbāh wa-l-Naẓāʾir, a legal digest. Tāj al-Dīn al-Subkī, al-Ashbāh wa-l-Naẓāʾir, ed. by Aḥmad ʿAbd al-Mawjūd and ʿAlī Muḥammad ʿIwaḍ, 2 vols. (Beirut: Dār al-Kutub al-ʿIlmīya, 1991).

Biographical Summary

Tāj al-Dīn al-Subkī was born and educated in Cairo, Egypt, in 1327. He was first educated by his father, the celebrated scholar Taqī al-Dīn al-Subkī, an influential figure in the umma. At age 11 years he joined his father in Damascus, where he studied under the leading scholars of his day, such as the historian al-Dhahabi and the faqih Ibn al-Naqīb. Aged 18 he became a mudarris professor) and khaṭīb at the Umayyad Mosque. In his late twenties he began to assist his father as qāḍī (Chief judge) of Syria, and on his father's retirement to Cairo in 1354, he replaced him as qāḍī of Damascus. He also held the title of Mufti.

In 1357 Tāj al-Dīn al-Subkī was removed from office but reinstated several months later. In 1368 he was jailed for misappropriation of funds. Following a petition by friends, he was released after 80 days and was exonerated.

Tāj al-Dīn al-Subkī died of the plague in 1370 at age of 44 years.

35. Al-Shatibi

Ibrahim bin Mosa bin Muhammad al-Shatibi al-Gharnati

(for short Abū Isḥāq Ibrāhīm ibn Mūsā al-Shāṭibī),

(1320 – 1388 C.E.),

was an Andalusí Sunni fiqih in the Maliki fiqh.

Scientific Contributions

Al-Shāṭibī learned from very prominent scholars of his time. He became a master in Arabic language and ittihad and research at a very early age. He would discuss various topics with his teachers before arriving to any conclusion.

Following are some works by al-Shāṭibī.

- Al-I'tiṣām (كتاب الاعتصام), 794696261 - This famous book of Imam Shatibi is the ultimate encyclopedia on the topic of defining religious innovations. It consists of 10 chapters. The introduction is written by Syed Rasheed Radha Al-Misri. This mammoth book was published by Dar al-Kutb Al-Arabiya in 1931 in Cairo.
- Al-Muwafaqaat fi Usool al-Sharia (الموافقات في اصول الشريعة), 37140768 - This is also one of Imam Shatibi's best known books. It is on the topic of Usul al-fiqh, and Islamic fiqh and Maqasid Al-Sharia (higher objectives). It was published by Dawlat Al-Tunisia in four volumes, translated and published into English as "The Reconciliation of the Fundamentals of Islamic Law".
- Shara ala al-Khutasa - This grammar book is about Ilm al-Nahw.
- Al-Itifaq fi Elm al-Ishtiqaq - This grammar book was on the topic of Ilm al-Sarf, or Arabic morphology, but it was lost during his life.
- Kitab al-Majalis - This book included commentary on Sahih Bukhari book al-Kitab Al-Biyooh.

- Kitab Al-ifidaat wa Al-inshadaat - This book included two volumes on Literature.

Following are some recent references to the work of al-Shāṭibī.

- Dr. Ahmad Raysuni, Imam Shatibi's Theory of the Higher Objectives and Intents of Islamic Law, translated by Nancy Roberts, publisher IIIT.
- Muhammad Khalid Masud, Islamic Legal Philosophy: A Study of Abu Ishaq al-Shatibi's Life and Thought, McGill University 1977.
- Wael B. Hallaq, A History of Islamic Legal Theories, Cambridge 1997, Ch. 5.
- The Shatibi Center, The Life of Al-Imam Ash-Shatibi.

Biographical Summary

Al-Shāṭibī's family descended from the Banu Lakhm. His Kuniyat was "Abu Ishaq", and his surnames were "Al-Lakhmi", "Al-Gharnati", "Al-Maliki" and "As-Shatibi". The date and place of his birth are unknown, around 1320 AD. However, one of his surnames, "As-Shatibi", points to the city Xàtiva, which indicates that he was a descendant of migrants from that town.

Mutakallimeen

Kalam is the science of research into nature of Allah, significance of the Names of Allah, significance of Quran and other scriptures, significance of Prophet Mohammad and other prophets, the significance of the angles, and the meaning and significance of the articles of Iman. As a result of discourses in Kalam, a rational concept emerges regarding the desirable attitude towards Allah and the Prophet in life, and the meaning and significance of Iman. This concept is called 'Aqeeda. As is clear from this discussion, 'Aqeeda is vaster than Iman as Iman is only a component of 'Aqeeda. Iman is described in Quran as commitment to the significance and role of Allah, Prophet, Scriptures, Angels, and the Resurrection on the Day of Judgement. Whereas, 'Aqeeda is not discussed in the Quran as a distinct concept.

A person who is skilled in Kalam is called Mutakallim whose plural is Mutakallimeen. We will use Kalam to represent the philosophical content of the religion; and we will use Mutakallim for the practitioner of kalam. These terms are not equivalent to the Christian terms, theology and theologian respectively.

36. Ali ibn al-Madini

Abū al-Ḥasan ʻAlī ibn ʻAbdillāh ibn Jaʻfar al-Madīnī
(Arabic: أبو الحسن علي بن عبد الله بن جعفر المديني),
(778 CE/161 AH – 849/234),
was a ninth-century alim who was influential in the science of hadith. Alongside Ahmad ibn Hanbal, Ibn Abi Shaybah and Yahya ibn Ma'in, Ibn al-Madini has been considered by many mohadditheen to be one of the four most significant authors in the field.

Scientific Contributions

Al-Nawawī said Ibn al-Madīnī authored approximately 200 works some on subjects not previously written about and many not since superseded. Following are some of the books.

- al-ʻIlal – on the subject of hidden defects (`ilal) in the sanads of hadith; of which a small segment has been published.
- Kitāb al-Ḍuʻafā' – on the subject of weak hadith narrators in the discipline of biographical evaluation.
- al-Mudallisūn – on the subject of hadith narrators who utilize ambiguous terminology in narrating.
- al-Asmā' wa al-Kunā – on names and epithets.
- al-Musnad – a collection of hadith arranged by narrators of ahadith.
- Kitab Ma'rifat al-Sahaba – The Book of Knowledge of the Companions.

Biographical Summary

Ibn al-Madīnī was born in the year 778 CE/161 AH in Basra, Iraq, to a family with roots in Medina, Hejaz. His teachers include his father, ʻAbdullāh ibn Jaʻfar, Ḥammād ibn Yazīd, Hushaym and Sufyān ibn ʻUyaynah and others from their era. His teacher, Ibn ʻUyaynah, said that he had learned more from Ibn al-Madīnī, his student, than his student from him.

Ibn al-Madīnī specialized in the disciplines of hadith, biographical evaluation and al-'Ilal, hidden defects, in the sanad, chain of narration. He was praised by other hadith specialists for his prowess in that field - by both his contemporaries, students and his teachers. 'Abd al-Raḥmān ibn Mahdī described Ibn al-Madīnī as the most knowledgeable person of prophetic hadith.

His students include prominent hadith scholars in their own right. They include: Muḥammad ibn Yaḥyā al-Dhuhalī, Muḥammad ibn Ismā'īl al-Bukhārī, Abū Dāwūd Sulaymān ibn al-Ash'ath al-Sijistānī and others. Al-Bukhārī, who went on to collect what is considered to be the most authentic collection of hadith in Sunni Islam, said that he did not consider himself diminutive in comparison to anyone other than Ibn al-Madīnī.

Al-Dhahabī lauded Ibn al-Madīnī as an imam and as exemplary to subsequent scholars in the field in hadith, a description he considered tarnished by Ibn al-Madīnī's adopted position in the inquisition of the ninth century. According to Al-Dhahabī, he adopted a position in favor of the Mu'tazilah regarding the uncreated origin of the Quran, but later regretted this and declared the claimants who hold that the Quran was created as apostates.

Ibn al-Madīnī died in Samarra, Iraq in June, 849.

37. Ibrahim al-Nazzam

Abū Isḥāq Ibrāhīm Ibn Sayyār Ibn Hāni‘ an-Naẓẓām

(Arabic: أبو إسحاق بن سيار بن هانئ النظام),

(c. 775 – c. 845),

was a Mu'tazilite mutakallim and poet. He was a nephew of the Mu'tazilite mutakallim Abu al-Hudhayl al-'Allaf, and al-Jahiz was one of his students. Al-Naẓẓām served at the courts of the Abbasid Caliph al-Mamun. His ilm-al-kalam doctrines and works are lost except for a few fragments.

Scientific Contributions

Ibrahim al-Nazzam diverged from many of the widely held views of his time.

- He was famous for his strong rejection of analogical reasoning: analogical reasoning was accepted by both the Hanafites and Shafi'ites.
- He rejected the idea of "Istihsan" which stands for the discretion that faqihs use to express their preference for particular judgements in Islamic law over other possibilities; it is one of the principles of legal thought underlying scholarly interpretation or "ijtihad".
- He rejected the doctrine of binding consensus (Ijma): used by all of Sunni Islam as a pillar of fiqh.
- He rejected the reports supposedly transmitting prophetic traditions as narrated by Abu Hurayra: these are accepted by jurists in all Sunni sects.

Like other early Mu'tazilites, al-Naẓẓām was a scripturalist who had no use for the traditions and accounts supposedly related by Abu Hurayra. However, Abu Hurayra is the most prolific ḥadīth narrater, who is accepted by all Sunni Jurists, even in spite of these narrations being full of incongruities, as argued by Ibrahim al-Nazzam.

For al-Naẓẓām, both of the so-called single-source and multiple-source reports, such as the multitudinous narratives variously attributed to Abu Hurayra, could not be trusted.

Al-Naẓẓām bolstered his refutation of the thitherto long-held esteem of the accounts of Abu Hurayra, and some others, using a larger claim. The claim asserts that such reports circulated and thrived mainly to support and legitimize the polemical causes of various theological sects and fuqaha.

The argument is a generalization of the common stance of the Sunni Jurists. This stance asserts that no single transmitter, be he contemporaneous with Muhammad or not, could by himself be held above suspicion of altering the content of any single report.

Al-Naẓẓām's skepticism involved excluded the possible verification of a report narrated by Abu Hurayra, whether it is traced back to a single source (wāḥid) or many (mutawātir). In addition, Ibrahim al-Nazzam also questioned other reports of widespread acceptance; these had proved pivotal to classical Muʿtazilite criteria devised for verifying the single reports.

Shi'ite mutakallemun al-Shaykh al-Mufid and Sharif al-Murtaza held in high esteem al-Naẓẓām's Kitāb al-Nakth (The Book of Dismantling), in which he denied the doctrinal validity of consensus (Ijma). Al-Naẓẓām's rejection of consensus was primarily due to his rationalist criticism of some of the first generation of Muslims, *whom he viewed as possessing defective personalities and intellects.*

Biographical Summary

Ibrahim al-Nazzam was born in 775 AD, and he died in 845 AD.

Naẓẓām spent his youth in Basra. There he studied *kalām* under the great Muʿtazilite mutakallim Abū al-Hudhayl al-ʿAllāf, his maternal uncle, but soon broke away from him.

He took part in scholarly debates, while in Basra, in his early youth. He moved to Baghdad in the early 820s, where he received the support of the Abbasid caliphs until his death in 845.

He taught many Mu‹tazilite ulama of the ninth century, among whom was his follower al-Jahiz.

38. Abu Mansur al-Baghdadi

Abū Manṣūr ʿAbd al-Qāhir ibn Ṭāhir bin Muḥammad bin ʿAbd Allāh al-Tamīmī al-Shāfiʿī al-Baghdādī

(Arabic: أبو منصور عبدالقاهر ابن طاهر بن محمد بن عبدالله التميمي الشافعي البغدادي),

(980 – 1037 AD)

was a Shafi'I alim, mutakallim, Usul Imam, heresiologist and mathematician.

Scientific Contributions

'Abd al-Qahir al-Baghdadi wrote several books including Kitāb Uṣūl al-Dīn, a systematic treatise, beginning with the nature of knowledge, creation, how the Creator is known, His attributes, etc. He also wrote al-Farq bayn al-Firaq which takes each sect separately, judges all from the standpoint of orthodoxy and condemns all which deviate from the straight path.

Both books were major works on the beliefs of Ahl al-Sunna.

He also wrote the treatise al-Takmila fi'l-Hisab which contains results in number theory, and comments on works by al-Khwarizmi which are now lost.

Biographical Summary

'Abd al-Qahir al-Baghdadi was born in 980 AD, and was raised in Baghdad. He was a member of the Arab tribe of Banu Tamim. He received his education in Nishabur and subsequently taught 17 subjects, including law, usul, arithmetic, law of inheritance and ilm-al-kalam. Most of the scholars of Khurasan were his pupils. Ibn 'Asakir writes that Abu Mansur met the companions of the companions of Imam al-Ashari and acquired knowledge from them.

'Abd al-Qahir al-Baghdadi died in Esfarayen in 1037 AD.

39. Abu al-Hasan al-Ash'ari

Abū al-Ḥasan ʿAlī ibn Ismāʿīl ibn Isḥāq al-Ashʿarī

For short Al-Ashʿarī (الأشعري),

(c. 874–936 (AH 260–324)),

was a mutakallim and eponymous founder of Ashʿarism or Asharite kalam.

Scientific Contributions

Al-Ashʿarī was notable for taking an intermediary position between the two diametrically opposed schools of kalam prevalent at the time.

The Muʿtazilites, advocated the use of reason in theological debate and believed the Quran was created, as opposed to uncreated Word of Allah. On the other hand, the Zahirites and Muhadditheen, were opposed to the use of philosophical reason (Kalam) and condemned any debate altogether.

Abu al-Hasan al-Ash'ari refuted the Muʿtazilites by stating "if the Quran was created then that implied God created this knowledge, and thus did not have knowledge of the Quran before this, and this contradicts God's omnipotence as he is all knowing, and therefore must have always had knowledge of the Quran". Al-Ashʿari's school eventually won "wide acceptance" within some sects of Sunni Islam. However, the Shi'a do not accept his beliefs, as Ashari's works involved refuting Shi'ism and Mu'tazilism, which was the doctrine held by Shi'as.

The original versions of this text by Abu al-Hasan al-Ash'ari did not survive.

After leaving the Muʿtazila school, and joining the side of traditionalist mutakallimeen al-Ash'ari formulated the kalam of Sunni Islam. He was followed in this by a large number of distinguished ulama, most of whom belonged to the Shafi'i school of fiqh. The most famous of these are Abul-

Hassan Al-Bahili, Abu Bakr Al-Baqillani, Al-Juwayni, Al-Razi and Al-Ghazali. Thus Al-Ash'ari's school became, together with the Maturidi, the main schools reflecting the beliefs of the Sunnah.

In line with Sunni tradition, al-Ash'ari held the view that a Muslim should not be considered an unbeliever on account of a sin even if it were an enormity such as drinking wine or theft. This opposed the position held by the Khawarij.

Al-Ash'ari also believed it impermissible to violently oppose a leader even if he were openly disobedient to the commands of the sacred law.

Al-Ash'ari spent much of his works opposing the views of the Muʿtazila school. In particular, he rebutted them for believing that the Qur'an was created and that deeds are done by people of their own accord. He also rebutted the Muʿtazili school for denying that Allah can hear, see and has speech. Al-Ash'ari confirmed all these attributes stating that they differ from the hearing, seeing and speech of creatures, including man.

He was also noted for his teachings on atomism.

The Salafis argue that Al-Ash'ari had accepted the Salafi theology before his death.

The Ashari alim Ibn Furak numbers Abu al-Hasan al-Ash'ari's works at 300, and the biographer Ibn Khallikan at 55; Ibn Asāker gives the titles of 93 of them, but only a handful of these works, in the fields of heresiography and kalam, have survived. Following are the three main ones:

- Maqalat al-Islamiyyin wa Ikhtilfa al-Musallin ("The Discourses of the Proponents of Islam and the Differences Among the Worshippers"), an encyclopaedia of deviated Islamic sects. It comprises not only an account of the Islamic sects but also an examination of problems in kalām, or religious discourses, and the

Names and Attributes of Allah; the greater part of this work seems to have been completed before his conversion from the Muʿtaziltes.

- Kitāb al-ibāna 'an usūl al-diyāna.
- Al-Luma`

  Al-Luma` fi-r-Radd `ala Ahl al-Zaygh wa al-Bida` ("The Sparks: A Refutation of Heretics and Innovators"), a slim volume.

  Al-Luma` as-Saghir ("The Minor Book of Sparks"), a preliminary to al-Luma` al-Kabir.

  Al-Luma` al-Kabir ("The Major Book of Sparks"), a preliminary to Idah al-Burhan and, together with the Luma` al-Saghir, the last work composed by al-Ash`ari according to Shaykh `Isa al-Humyari.

The 18th century Islamic scholar Shah Waliullah regarded Al-Ash'ari as a Mujaddad. According to him: A Mujadid appears at the end of every century: The Mujadid of the first century was Imam of Ahlul Sunnah, Umar bin Abdul Aziz. The Mujadid of the second century was Imam of Ahlul Sunnah Muhammad Idrees Shaafi. The Mujadid of the third century was the Imam of Ahlul Sunnah, Abu al-Hasan al-Ash'ari. The Mujadid of the fourth century was Abu Abdullah Hakim Nishapuri.

Earlier major scholars also held positive views of al-Ash'ari and his efforts, among them Qadhi Iyad and Taj al-Din al-Subki.

Although the Ash'ari school of theology is often called the Sunni 'orthodoxy,' the original ahl al-hadith, early Sunni creed from which Ash'arism evolved, has continued to thrive alongside it as a rival Sunni 'orthodoxy' as well.

Biographical Summary

Al-Ash'ari was born in Basra, Iraq, and was a descendant of the prophet Muhammad's companion, Abu Musa al-Ashari. As a young man he studied under al-Jubba'i, a renowned teacher of Muʿtazilite theology and philosophy.

He remained a Muʿtazalite until his fortieth year when al-Ash'ari saw Muhammad in a dream three times in Ramadan. The first time, Muhammad told him to support what was related from himself, that is, the traditions (hadiths). Al-Ash'ari became worried as he had numerous strong proofs contradictory to the traditions.

After 10 days, he saw Muhammad again: Muhammad reiterated that he should support the traditions. So, Al-Ash'ari forsook Kalam and started following the traditions alone.

On the 27th night of Ramadan, he saw Muhammad for the last time. Muhammad told him that he had not commanded him to forsake Kalam, he had only told him to support the traditions narrated from himself. Thereupon Al-Ash'ari started to advocate the Hadith, finding proofs for these that he said he had not read in any books.

After this experience, he left the Muʿtazalites and became one of its most distinguished opponents, using the philosophical methods he had learned. Al-Ash'ari then spent the remaining years of his life engaged in developing his views and in composing polemics and arguments against his former Muʿtazalite colleagues. He is said to have written up to three hundred works, of which only four or five are known to be extant.

Al-Ash'ari died in Baghdad in 936 AD.

40. Al-Shaykh Al-Mufid

Abu 'Abd Allah Muhammad ibn Muhammad ibn al-Nu'man al-'Ukbari al-Baghdadi,

(known as al-Shaykh al-Mufid),

(Arabic: الشيخ المفيد),

(c. 948–1022 CE),

was a prominent mutakallim.

Scientific Contributions

Shaykh al-Mufid is said to have written 200 works, of which only a few more than ten have survived. Some of his works are as follows:

- Al-Amali (of Shaykh Mufid), also known as "Al-Majaalis", traditions recorded by al-Mufid's pupils during the sessions where al-Mufid gave the chain of narration ending up with himself.
- Tashih al-Itiqadat, a correction of al-Saduq's Risalat al-Itiqadat
- Awail Al Maqalat, an elaboration of al-Mufid's theology and "a practical catalogue of Imamite positions on disputed questions".
- Kitab al-Irshad or Al-Irshad fi ma'rifat hujaj Allah 'ala al-'ibad, on the lives of the Shia Imams.
- Al-Fusul al-`Ashara fi al-Ghaybah.
- Ahkam al-Nisa, on legal obligations regarding women.
- Fifth Risalah on Ghaybah.
- Al-Muqni'ah (The Legally Sufficient) The commentary on this book by Shaykh Tusi, Tadhhib al-Ahkam fi Sharh al-Muqni'ah, is among the Shia four books.

Biographical Summary

Al-Mufid was born in 'Ukbara, a small town to the north of Baghdad, on 11th Dhul Qa'dah in 336 Hijra. According to Shaykh Tusi, however, he was born in 338 AH. He was also called Ibn Muallim, meaning "son of the teacher"; his father was a Muallim (teacher).

Later he migrated with his father to Baghdad, where the Shia Buwayhids were ruling. He studied with Ibn Babawayh.

Sharif al-Murtaza and Shaykh Tusi were among his students. His career coincided with that of the Mu'tazili mutakallim and leader of the Bahshamiyya school, 'Abd al-Jabbar.

Al-Mufid was often attacked, and his library and school were destroyed.

Among his teachers were the mutakallim Abu Ali al-Iskafi, Abu Abdallah al-Marzubani, Abu Abdallah al-Basri, Abu al-Hassan, and Ali ibn Isa al-Rummani. He was taught the Islamic science of hadith by Al-Shaykh al-Saduq.

Al-Mufid is regarded as the most famous scholar of the Buyid period and an eminent faqih, mainly due to his contributions in the field of kalam. According to Ibn al-Nadim, who knew al-Mufid personally, he was the head of the Mutekallimeen in the field of kalam. Al-Tawhidi, who was also personally familiar with al-Mufid, described him as "eloquent and skillful at dialectic (jadal)". His skill in polemical debate was such that he was said to be capable of convincing his opponents "that a wooden column was actually gold".

Al-Mufid died in 1022 AD. Al-Mufid died on the third day of Ramadan in 413 AH. He remained buried in his own house for two years, after which his body was moved to Al Kadhimiya Mosque and buried next to his teacher, Ibn Qulawayh al-Qummi.

41. Abu Hamid Muhammad Al-Tusi Al-Ghazali

Abu Hamid Muhammad ibn Muhammad al-Tusi al-Ghazali

archaically Latinized as Algazelus,

(in Arabic أَبُو حَامِد مُحَمَّد بْن مُحَمَّد الطُّوسِيّ الغَزَالِي),

(c. 1058 –19 December 1111),

was a prominent and influential faqih, mutakallim, mufti, theologian, mystic, (anti)philosopher, and logician.

Scientific Contributions

Al-Ghazali is considered to be the 11th century's Mujaddid, a renewer of Muslim faith, who appears once in 100 years to restore the faith of the Islamic community, as believed by some Muslim circles.

Al-Ghazali was a prominent faqih in the Shafi'i school of law. He is referred to as Ḥujjat al-Islām in some Muslim circles, which means a case for Islam.

He traveled to Baghdad and progressed to the appointment as the head of the Nizamiyya Madrassah in Baghdad. It was a prestigious academic position. It is reported without much detail that he got disappointed, and resigned his position at Nizamiyya. He lived for about 10 years, away from lime light. The reasons are not known, but some speculate that he realized that he chose the path of status and ego over God. However, while he worked to bring Sufis and Sunni Muslims closer, it is not clear if he lived the non worldly and austere life of the Sufis. Many of his great works were written during this period: he wrote his magnum opus titled Iḥyā' 'ulūm ad-dīn ("The Revival of the Religious Knowledge"). The Tahāfut al-Falāsifa ("Incoherence of the Philosophy") is also his landmark work with respect to its impact, and some argue that it was a negative one, upon the Muslim attitude towards intellect and reason, as it advances a critique of science.

Al-Ghazali contributed significantly to the development of a systematic view of Sufism and its integration and acceptance in mainstream Islam. He belonged to the Shafi'i fiqh and to the Asharite school of kalam. He is viewed as a key member of the influential Asharite school of kalam and the most important refuter of the Mutazilite school of kalam. However, he chose a slightly different position with respect to beliefs and thoughts which differed in some aspects from the orthodox Asharite school.

A total of about 70 works can be attributed to al-Ghazali. He is also known to have written a fatwa against the Taifa kings of al-Andalus, declaring them to be unprincipled, not fit to rule and that they should be removed from power. This fatwa was used by Yusuf ibn Tashfin to justify his conquest of al-Andalus. This is an example of how the nexus between Muslim ulama and Muslim rulers has often worked in the past, and still works today.

Al-Ghazali's 11th-century book titled *Tahāfut al-Falāsifa* ("Incoherence of Philosophy") marked a major turn in Islamic epistemology. The encounter with skepticism led al-Ghazali to investigate a form of theological occasionalism, or the belief that all causal events and interactions are not the product of material conjunctions but rather the immediate and present will of God.

This is likely connected with the reported crisis that Al-Ghazali went though around 1095, when he was around 37 years of age. This crisis seems to have shaken him up, leading him to leave Baghdad and start his teachings in Tus, where he himself built the facilities and operated them as well. He was going through a crisis, but he also had a powerful capacity to write convincingly. It unfortunately led to works like *Tahāfut al-Falāsifa,* which weakened the scientific mindset that Muslims had developed from the days of the Prophet, and possibly inflicted the thinking of Muslim Ulama and Muslim masses with similar crisis that he himself was going through.

Ibn Rushd (1126-1198) wrote a detailed rebuttal of al-Ghazali's *Tahāfut al-Falāsifa.* However, during the decadės that had passed, the epistemological course of Islamic thought had already been set along lines that greatly deviated from what Muslims had developed since the days of the Prophet. Al-Ghazali gave an example of the illusion, as he claimed it to be, of scientific laws of cause and effect. He considered the fact that cotton burns when coming into contact with fire. While it might seem as though a natural law was at work, it happened, al-Ghazali claimed, only because God willed it to happen each and every time: the event was "a direct product of divine intervention versus any more attention-grabbing science". This view is called al-Ghazali's theological occasionalism. Averroes, by contrast insisted that while God created the natural law, humans "could more usefully say that fire caused cotton to burn because creation had a pattern that they could discern".

The *Tahāfut al-Falāsifa* marked a turning point in Muslim thought process in its vehement rejections of science and rationality. The book took aim at the philosophy and rationalism itself, versus the thought process by the earlier scientists like Avicenna and al-Farabi.

Professor of Mathematics Nuh Aydin wrote in 2012 that one of the most important reasons for the decline of science in the Islamic world has been Al-Ghazali's attack on *philosophers* (scientists, physicists, Mathematicians, logicians). The attack peaked in his book *Tahāfut al-Falāsifa* whose central idea of theological occasionalism implies that *philosophers* cannot give rational explanations to either metaphysical or physical questions. The idea caught on with the religious Ulama, and through them with the general Muslim masses, and nullified the critical thinking in the Islamic world.

Another of al-Ghazali's major works is *Ihya' Ulum al-Din* or *Ihya'u Ulumiddin* (*The Revival of Religious Knowledge*). It covers almost all fields of Islamic knowledge: fiqh, kalam, and sufism.

It contains four major sections: *Acts of worship* (*Rub' al-'ibadat*), *Norms of Daily Life* (*Rub' al-'adatat*), *The ways to Perdition* (*Rub' al-muhlikat*) and *The Ways to Salvation* (*Rub' al-munjiyat*). Some circles among Muslims say that the *Ihya* became the most frequently recited Islamic text after the Qur'an and the hadith. Its effect was to uplift orthodoxy in Islamic theology, as a comprehensive guide to every aspect of Muslim life and death. It also tried to bring Sufi mysticism into mainstream Islamic Theology. The book was well received by the class of Islamic Ulama such as Nawawi who grossly over stated that: "Were the books of Islam all to be lost, excepting only the Ihya', it would suffice to replace them all." This reception, however, was not universal as the book was burned in Almoravid Spain in 1109 and 1143. Al-Ghazali criticized the fuqaha for meddling in politics, and this despite al-Ghazali's own fatwa against Taifa kings of al-Andalus. Al-Ghazali was also criticized due to his syncretism and support of Sufism.

A rewritten version of Ihya is the book titled *The Alchemy of Happiness.*

One of the key sections of Ghazali's *Revival of the Religious Knowledge* is *Disciplining the Soul*, which focuses on the internal struggles that al-Ghazali opined every Muslim will face over the course of his lifetime. This may be an example of how Al-Ghazali unknowingly inflicts his own crisis on all Muslims.

One chapter primarily focuses on how one can develop himself into a person with positive attributes and good personal characteristics, which aspect within him had presumably been impaired by his own crisis. Another chapter focuses on sexual satisfaction and gluttony. Here, Ghazali presumes that indeed every man has these desires and needs, and that it is natural to want these things. Al-Ghazali claims this despite the Prophet having explicitly stated that there is a middle ground for man in order to practice the tenets of Islam faithfully. These two chapters were the 22nd and 23rd chapters, respectively, in Ghazali's *Ihya*.

Al-Ghazali crafted his rebuttal of the scientific viewpoint in his book on the creation of the world titled *The Eternity of the World*. Al-Ghazali essentially formulates an argument for what he views as a sacrilegious thought process. Al-Ghazali represents scientific approach as the concept that motion will always precede motion, or in other words, a force will always create another force, and therefore for a force to be created, another force must act upon that force. This means that in essence time stretches infinitely both into the future and into the past, which means that God did not create the universe at one specific point in time. Al-Ghazali then counters this by first stating that if the world was created with exact boundaries, then in its current form there would be no need for a time before the creation of the world by God.

In his work on *The Decisive Criterion for Distinguishing Islam from Clandestine Unbelief Al*-Ghazali lays out his approach. Ghazali veers from the often-hardline stance of many of his contemporaries during his time period and states that as long as one believes in the Prophet Muhammad and God Himself, there are many different ways to practice Islam and that any of the many traditions practiced in good faith by believers should not be viewed as heretical by other Muslims. While Ghazali does state that any Muslim practicing Islam in good faith is not guilty of apostasy, he does outline in this book that there is one standard of Islam that is more correct than the others, and that those practicing the faith incorrectly should be moved to change. In Ghazali's view, only the Prophet himself could deem a faithfully practicing Muslim an infidel, and his work was a reaction to the religious persecution and strife that occurred often during this time period between various Islamic sects. Unfortunately, sectarian strife still exists among Muslims.

Al-Ghazali wrote an autobiography towards the end of his life, titled *Deliverance From Error* (المنقذ من الضلال *al-Munqidh min al-Dalal*). In it, al-Ghazali recounts how, once a crisis of epistemological skepticism had been resolved by "a light which God Most High cast into my breast ...". This he says is the key to most knowledge that he studied and mastered, including the arguments of kalam, Islamic philosophy, and Ismailism. Though appreciating what was valid in the first two of these, at least, he determined that all three approaches were inadequate and he found ultimate value only in the mystical experience and insight he attained as a result of following Sufi practices. Such observations which al-Ghazali presents and argues, presumably represent his own personal journey out of the darkness of his own crisis.

Al-Ghazali wrote most of his works in Persian and in Arabic. His most important Persian work is *Kimiya-yi sa'adat* (The Alchemy of Happiness). It is al-Ghazali's own Persian version of *Ihya' 'ulum al-din* (The Revival of Religious Knowledge) in Arabic, but a shorter work. The book was published several times in Tehran by the edition of Hussain Khadev-jam, a renowned Iranian researcher. It is translated to English, Arabic, Turkish, Urdu, Azerbaijani and other languages.

Another authentic work of al-Ghazali is the so-called "first part" of the Nasihat al-muluk (Counsel for kings), addressed to the Saljuq ruler of Khurasan, Ahmad b. Malik-shah Sanjar (r. 490-552/1097-1157). The text was written after an official reception at Sanjar court in 503/1109. Al-Ghazali was summoned to Sanjar because of the intrigues of his opponents and their criticism of his student's compilation in Arabic, al-Mankhul min taʿliqat al-usul (The sifted notes on the fundamentals); an additional concern, presumably, was his refusal to continue teaching at the Nizamiya of Nishapur. After the reception, al-Ghazali had, apparently, a private audience with Sanjar, during which he quoted a verse from the Quran 14:24: "Have you not seen how Allah sets forth a parable of a beautiful phrase (being) like a beautiful tree, whose roots are firm and whose branches are in Heaven."

The text of the "Nasihat al-muluk" is actually an epistle with a short explanatory note on "al-Mankhul min taʿliqat al-usul" added on its frontispiece.

The majority of other Persian texts, ascribed to him with the use of his fame and authority, especially in the genre of Mirrors for Princes, are either deliberate forgeries fabricated with different purposes or compilations falsely attributed to him. The most famous among them is Ay farzand (O Child!). This is undoubtedly a literary forgery fabricated in Persian one or two generations after al-Ghazali's death. The sources used for the forgery consist of two genuine letters by al-Ghazali (number 4, in part, and number 33, totally); both appear in the *Faza'il al-anam.* Another source is a letter known as *Ayniya* and written by al-Ghazali's younger brother Majd al-Din Ahmad al-Ghazali (d. 520/1126) to his famous disciple ʿAyn al-Quzat Hamadani (492-526/1098-1131); the letter was published in the *Majmuʿa-yi athar-i farsi-yi Ahmad-i Ghazali* (Collection of the Persian writings of Ahmad Ghazali). The other is ʿAyn al-Quzat's own letter, published in the *Namaha-yi ʿAyn al-Quzat Hamadani* (Letters by ʿAyn al-Quzat Hamadani). Later, *Ay farzand* was translated into Arabic and became famous as *Ayyuha al-walad*, the Arabic equivalent of the Persian title. The earliest manuscripts with the Arabic translation date from the second half of the 16th and most of the others from the 17th century. The earliest known secondary translation from Arabic into Ottoman Turkish was done in 983/1575. In modern times, the text was translated from Arabic into many European languages and published innumerable times in Turkey as Eyyühe'l-Veled or Ey Oğul.

A less famous "Pand-Nama" (Book of counsel) also written in the genre of advice literature is a very late compilatory letter of an unknown author formally addressed to some ruler and falsely attributed to al-Ghazali, obviously because it consists of many fragments borrowed from various parts of the Kimiya-yi saʿadat.

During his life, Al-Ghazali wrote over 70 books on knowledge, Islamic reasoning and Sufism. Al-Ghazali tried to bring together Sufism with Shariah. However, they remained apart.

In his works he made an effort to present a formal description of Sufism. However, today no such definition is generally accepted.

His works strengthened the status of Sunni Islam against other schools. The Batininiyya (Ismailism) had emerged in Persian territories and were gaining power during al-Ghazali's period, as Nizam al-Mulk was assassinated by the members of Ismailis. In his *Fada'ih al-Batiniyya* (*The Infamies of the Esotericists*) al-Ghazali declared them unbelievers whose blood may be spilled. Al-Ghazali worked for the general acceptance for Sufism, at the expense of philosophy. At the same time, in his refutation of philosophers he made use of their philosophical categories and inadvertently helped to give them circulation.

The staple of his religious philosophy was to argue that the Creator was the center point that played a direct role in all world affairs and in all human life. This view is referred to as al-Ghazali's Occasionalism Theology.

One of the efforts of al-Ghazali was his writing and reform of education; it laid out the path of Islamic Education that was used from the 12th to the 19th centuries. However, his approach to refute scientific theories and his insistence to offer explanation of physical phenomena invoking direct divine intervention, produced an educational system that was arguably backward. Similar education still is in use in some places, and upholds ideas like a Geocentric planetary system.

Al-Ghazali mentioned the number of his works "more than 70" in one of his letters to Sultan Sanjar in the late years of his life. Some "five dozen" are plausibly identifiable, and several hundred attributed works, many of them duplicates because of varying titles, are doubtful or spurious.

In Deliverance from Error, Ghazali states that religion and science (philosophy in his parlance) do not deal with the same subjects; and to condemn the study of science for fear that it endangers religion is to mistake the place of each of them, implying a separation between religion and science, an idea that deviated from early Muslim practices. He writes that the books of the philosophers be banned – (he defines philosophy as composed of six branches: mathematical, logical, physical, metaphysical, political, and morale).

Biographical Summary

Al-Ghazali was born in 1058 in Tabaran, a town in the district of Tus in then Khorasan province of Iran. It is not understood why he is called al-Ghazali.

It was the time when the Seljuks entered Baghdad and ended Shia Buyid Amir al-Umaras. This marked the start of Seljuk influence over Caliphate. Influence was maneuvered in multiple ways; for example, Abu Suleiman Dawud Chaghri Beg married his daughter, Arslan Khatun Khadija, to caliph al-Qa'im in 1056.

Al-Ghazali's contemporary and first biographer, 'Abd al-Ghafir al-Farisi, records that al-Ghazali began to receive instruction in fiqh from Ahmad al-Radhakani, a local teacher, and from Abu Ali Farmadi, a Naqshbandi Sufi from Tus. He later studied under al-Juwayni, the distinguished faqih and mutakallim, and an outstanding teacher in Nishapur. After al-Juwayni's death in 1085, al-Ghazali departed from Nishapur and joined the court of Nizam al-Mulk, the powerful vizier of the Seljuk empire, which was likely seated in Isfahan. Nizam al-Mulk advanced al-Ghazali's career by appointing him in July 1091 to the prestigious position of director at the Nizamiyya madrassa in Baghdad.

He underwent a crisis in 1095, which some speculate was brought on by clinical hysteria, while others speculate that it was a spiritual thing. He abandoned his career at Nizamiyya, and left Baghdad. He made arrangements for his family, and started for Haj. After some time in Damascus and Jerusalem, with a visit to Medina and Mecca in 1096, he returned to his home place in Tus where he stayed away from limelight for about a decade. He continued to publish, receive visitors and teach in the zawiya (a Sufi meditation place) and khanqah (a Sufi lodge). Al-Ghazali built and operated these facilities himself.

Fakhr al-Mulk, grand vizier to Ahmad Sanjar, pressed al-Ghazali to direct the Nizamiyya madrassah in Nishapur. Al-Ghazali accepted it in 1106, reportedly reluctantly, fearing that he and his teachings would meet with resistance and controversy. He later returned to Tus and declined an invitation in 1110 from the grand vizier of the Seljuq Sultan Muhammad I to return to Baghdad.

He died on 19 December 1111. According to 'Abd al-Ghafir al-Farisi, he had several daughters but no sons.

Awwalagists

The Quran was revealed among a particular segment of humanity and at a particular historical epoch in the world. As Islam reached extend-ed regions of the world among different segments of humanity, and the time flowed on to new epochs in human history, it became necessary to reevaluate the message of Quran for these new segments of humanity and for the new epochs in human flow. This is an obvious need not just for the Arabs, who were the original addressees of the Quran, but also internationally and globally. It is a very divisive, and sometimes even violent, question among Muslims as to how this task shall be performed.

Generally, there are two processes to do this task. They use what is called Tafsir and Ta'wil. While all Muslims acknowledge the need for Tafsir, a significant segment among Muslims does not acknowledge the need for Ta'wil, and some will even question if it makes sense to do Ta'wil. This circumstance exists notwithstanding the fact that Muslims do not agree upon definitions and scopes of these two processes. Some of the confusion is created by the Orientalists who use limiting and misleading terms in related discourses. For example, they generally use the term 'traditionalist' for people of and practitioners of hadith. Among many other issues, the use of the word 'traditionalist' produces much ambiguity and misinformation simply from the dictionary meaning of the word.

The process of Tafsir uses mainly the ahadith in order to explain the message of Quran. The process is sometimes limited by the requirement that any explanation of the Quran be compliant with, and to some extent subservient to, the work of Tafsir that has already been done by the past Mufassireen (people of Tafsir, with Mufassir its singular). This is especially so with respect to the Tafsir presented by the Ashab (apostils) of the

Prophet, and those who immediately followed the Ashab, known as Tabi'in. However, the restrictive requirement is applied even for more recent works of Tafsir, that a coherent and persistent community of Muslims, known as Ulama, have decided to uphold with their approval.

The circumstance makes the task of doing Tafsir insurmountably resistant for a newcomer. And we have not even begun to talk about Ta'wil about which there is not even a faint agreement among Muslims.

Given the situation with Tafsir and Ta'wil, we have decided to use a fresh concept that we have called Awwalogy. The term Awwalagists represents scientists who perform analysis to bring the textual meanings back to their roots, the original meaning intended by the text.

The term is intended to include both tafsir and ta'wil, as well as additional research. Tafsir means to explain, to expound, to elucidate, and to interpret. Ta'wil means to return, to revert, thereby returning to the original meaning of a word to understand its connotations.

This is not the same as exegetes because the term exegesis applies heavily to the Bible paying little heed to the process for other scriptures and languages. Also, the term exegetes is not rich enough to discriminate between tafsir and ta'wil, for instance. Further, the term exegesis is used also in broad connotations such as politics and literature. Therefore, the term Awwalagists is coined to represent the concept of research in Ta'wil and Tafsir.

The term Awwalagist carries a meaning that is fresh even for Muslim tradition. It is distinct from the term mufassir which carries quite a historical baggage and is entrenched in sectarian controversy. Independent of such historical baggage and controversy, the term Awwalagist carries a spirit of research in pursuit of the truth, irrespective of who the researcher is in the conventional sectarian landscape.

There is no concept of Awwalogy in English language, nor in Christian religious tradition where the integrity of religion and its continuity is delegated to Ecumenical Councils and the institute of Papacy.

42. Mujahid ibn Jabr

Mujahid ibn Jabr

(Arabic: مُجَاهِدُ بْنُ جَبْرٍ),

(645–722 CE),

was a Tabi'in and one of the major early ulama.

Scientific Contributions

It is related by Ibn Sa'd in the Tabaqat (6:9) and elsewhere that he went over the explanation of the Qur'an together with Ibn 'Abbas thirty times.

Mujahid ibn Jabr is said to be relied upon in terms of tafsir according to Sufyan al-Thawri, who said: "If you get Mujahid's tafsir, it is enough for you."

He is the first to compile a written tafsir of the Qur'an. His tafsir in general followed these four principles:

1. That the Qur'an can be explained by other parts of the Qur'an. For example, in his interpretation of Q 29:13, he refers to Q 16:25,
2. Interpretation according to traditions,
3. Reason,
4. Literary comments.

Al-Tabari's Jami' al-bayan attributes a significant amount of tafsir material to Mujahid.

Biographical Summary

Mujahid ibn Jabr was born in 645 AD and died in 722.

Mujahid ibn Jabr was one of the leading Qur'anic mofassirs of the generation after that of the Prophet Muhammad and his Companions. He is the first to compile a written tafsir of the Qur'an, in which he stated

> "It is not permissible for one who holds faith in Allah and the Day of Judgment to speak on the Qur'an without learning classical Arabic."

He is said to have studied under Amir al-Mu'minin 'Ali ibn Abi Talib until his martyrdom. At that point, he began to study under Ibn Abbas, a companion of the Prophet known as the father of Qur'anic tafsir. Mujahid ibn Jabr was known to be willing to go to great lengths to discover the true meaning of a verse in the Qur'an, and was considered to be a well-travelled man. However, there is no evidence he ever journeyed outside of the Arabian Peninsula.

43. Sahl al-Tustari

Abū Muḥammad Sahl ibn ʿAbd Allāh al-Tustarī

(Sahl al-Tustarī for short, (Arabic: سهل التستري)),

(also known as Sahl Shushtarī (Arabic: سهل شوشترى)),

(c.818 CE (203 AH) – c.896 CE (283 AH)),

was a an alim and early classical Sufi mystic. He founded the Salimiyah kalam school, which was named after his disciple Muhammad ibn Salim. Tustari is known for his Tafsir, a commentary on and interpretation of the Qur'an.

Scientific Contributions

From an early age he led a life of zuhd, with frequent fasting and study of the Qur'an and Hadith. He practiced repentance (tawbah) and, above all, constant remembrance of God (dhikr). This eventually culminated in a direct and intimate rapport with God.

Tustari was under the direction of the Sufi saint Dhul-Nun al-Misri for a time, and Tustari in turn was one of the teachers of Sufi mystic Mansur Al-Hallaj.

Tustari was close to the mohaddith and ravi Abu Dawood. He requested "O Abu Dawud, I want something from you." He said, "What is it?", to which Sahl replied, "On a condition that you say that you will fulfill it if possible." Abu Dawud replied in the affirmative. Sahl said, "Get out your tongue with which you narrated the hadiths of the Prophet (peace be upon him) so that I kiss it." Abu Dawud accept that and Sahl kissed his tongue. This shows the close affinity of early mohadditheen and sufis.

In these early days when the Sufis were becoming established, mostly in Baghdad; the notable Sufis of the time elsewhere were: Tustari in southwestern Iran, Al-Tirmidhi in Central Asia and the Malamatiyya.

Tustari commented on and interpreted the Qur'an. He maintained that the Qur'an "contained several levels of meaning". These included the outer or Zahiri meaning, and the inner or batini meaning. He also stated that "as a mystery of realization at the center of a sufi's personality, called the sir ('the secret'); and that the heart is where existence joins Being."

Tustari also "was the first to put the Sufi zikr of God on a firm theoretical basis. Tustari asserted that ultimately [...] it became clear to the zakir that the true agent of zikr was not the believer engaged in zikr but God Himself, who commemorated Himself in the heart of the believer. This realization of God's control over the heart led the believer to the state of complete trust in the Divine.

A recent translation of this awalagy is as follows:

ISBN 978-1-891785-19-1: Al-Tustari, Sahl ibn 'Abd Allah, December 2009, edited by Meri, Yousef: "Tafsir Al-Tustari: Great Commentaries of the Holy Qur'an". Translated by Keeler, Annabel; Keeler, Ali; and Fons Vitae.

Following are some of the statements attributed to al-Tustari.

- "I am the Proof of God for the created beings, and I am a proof for the awliya of my time".
- Asked "What is food?" Tustari replied: "Food is contemplation of the Living One."
- "Whoever wakes up worrying about what he will eat -- shun him!"
- "If any one shuts his eye to God for a single moment, he will never be rightly guided all his life long."

Bibliographic Summary

Sahl al-Tustari was born in the fortress town of Tustar (Arabic) or Shushtar (Persian) during the Abbasid Caliphate, in Khūzestān Province in what is now southwestern Iran.

44. Al-Qadhi al-Nu'man

Abū Ḥanīfa al-Nuʿmān ibn Muḥammad ibn Manṣūr ibn Aḥmad ibn Ḥayyūn al-Tamīmiyy

(Arabic: النعمان بن محمد بن منصور بن أحمد بن حيون التميمي),

generally known as al-Qāḍī al-Nu'mān (القاضي النعمان)

or as ibn Ḥayyūn (ابن حيون),

(died 974 CE/363 AH),

was an Isma'ili jurist and the official historian of the Fatimid Caliphate. He was also called Qaḍi al-Quḍāt (قَاضِي القضاة) "Jurist of the Jurists" and Dā'ī al-Du'āt (داعي الدعاة).

Scientific Contributions

Al-Qāḍī al-Nu'mān's work consists of over 40 treatises on fiqh, history, religious beliefs and Quranic esoteric tafsir. Fuat Sezgin cites 22 works by him.

Al-Qāḍī al-Nu'mān's most prominent work, the Da'a'im al-Islam (Arabic: دعائم الاسلام), which took nearly thirty years to complete, is an exposition of Isma'ili fiqh. This work was finally completed during the reign of Al-Mu'izz li-Din Allah (r. 953-975 CE/ 341-365 AH). It was accepted in its time as the official code of the Fatimid Caliphate. Up to this day it serves as the primary source of religious law (sharia) for some Musta'li communities, particularly Tayyibi Isma'ilis.

Iran incorporated the Da'a'im al-Islam into their constitution.

The book consists of 32 chapters in two volumes. The first volume consists of 7 chapters discussing the Seven pillars of Ismailism. The second volume consists of 25 chapters about various topics relating to different facets of life.

Foundation of Symbolic Interpretation (Asās al-Taʾwīl), is another one of Al-Qāḍī al-Nuʿmān's most celebrated works and deals with esoteric interpretation (taʾwīl). In the author's own words,

> "Our aim [with the Asās al-Taʾwīl] is to explain the inner dimension (bāṭin) of what we laid out in the book Daʿāʾim al-Islām, so that this book may be a source for the inner meaning (bāṭin), just as that one is for the external form (ẓāhir).

Al-Qāḍī al-Nuʿmān believed that it is important to recognize and understand the symbolism behind the stories in the Quran based on certain verses that point to an inner meaning such as:

> "Thus, your Lord will choose you and teach you the symbolic interpretation (taʾwīl) of events (aḥadīth)" (Quran 12:21).

In his work Foundation of Symbolic Interpretation (Asās al-Taʾwīl), he indicates that God made the Quran the miracle of Prophet Muhammad and its inner meaning the miracle of the Imams. Just as nobody can replicate the miracle of the Book, nobody can produce its inner meanings except for the Imams. This sacred knowledge is passed down through generations in their lineage and is entrusted to them.

Another major work of Al-Qāḍī al-Nuʿmān is the Kitab iftitah al-daʿwa wa-ibtida' al-dawla ("The Beginning of the Mission and Establishment of the State"). It narrates the rise of the Fatimids. It mentions the initial stages of the Isma'ili dawah in Yemen under Ibn Hawshab. It also discusses Abu Abdallah al-Shi'i's correspondence with the Kutama Imazighen and their military expeditions, leading to the conquest of the Aghlabids, who ruled Ifriqiya. It discusses Imam al-Mahdi's emigration from Salamiyah, his captivity in Sijilmasa and eventual release, culminating in the establishment of the Fatimid state in 909.

The book also gives an account of the circumstances leading to the revolt of al-Shi'i, for which it holds responsible the incitement of his elder brother Abu al-Abbas, and his later execution. It also gives a description of the Fatimid state up to the year 957, when the book was completed.

Ikhtilaf usul al-madhahib ("Differences Among the Schools of Fiqh") was a refutation of Sunni principles of Islamic Fiqh written at roughly the same time as the earliest of such works. Al-Qāḍī al-Nu'mān's book borrows heavily from those of Dawud al-Zahiri, Muhammad bin Dawud al-Zahiri and al-Tabari, three Sunni authors about whom Al-Qāḍī al-Nu'mān displays complex mixed feelings. It has been noted that while Al-Qāḍī al-Nu'mān's book is famous, it was not the first Ismali refutation of Sunni juristic principles.

Two other works of Al-Qāḍī al-Nu'mān are the Kitab al-majalis wa'l-musayarat ("The Book of Sessions and Excursions"), in which he has entailed in detail words of Imams in majlis, or just while walking which he had taken note of, and the wisdom encased within them); and the Kitab al-himma fi adab atba' al-a'imma ("The Book of Etiquette Necessary for Followers of the Imams").

Biographical Summary

Born in Kairouan, in what is now Tunisia, al-Qāḍī al-Nu'mān converted to Isma'ilism and began his career in Ifriqiya (now Tunisia, western Libya and eastern Algeria) under the first Fatimid caliph, Abdullah al-Mahdi Billah (r. 909-934 CE/297-322 AH). He quickly arose to become the most prominent qadhi of the Fatimid state.

His father, Muhammad ibn Mansur (d. 351 H/923 CE), had trained as a Sunni Maliki faqih in Qayrawan.

During his lifetime, Al-Qadhi al-Nu'man served four Fatimid Caliphs: Abdullah al-Mahdi Billah, al-Qa'im, al-Mansur Billah, and al-Mu'izz li-Din Allah.

The career of al-Qadhi al-Numan (d. 974) began under Al-Mahdi, the founder of Ismaili law and author of its most authoritative compendium, the Kitab da'a'im Al-Islam (Book of the pillars of Islam). In the absence of an Ismaili legal tradition, al-Qadhi Al-Numan relied primarily on the legal teaching of Imams Muhammad Al-Baqir and Jafar Al-Sadiq, transmitted by Twelver Shii traditionists, and secondarily on Zaydi traditions.

For example, Kitab da'a'im Al-Islam gives the imam authority for determining the beginning of the month without regard to the sighting of the new moon as required by all other Muslim legal schools. Since the early Fatimid period the beginning of the months was generally established in practice *on the basis of astronomical calculation* and thus often fell one or two days earlier than for other Muslims; this discrepancy often caused intercommunal quarrels about the beginning and end of the fasting month of Ramadan.

During al-Nu'man's fifty years of service to the Fatimids, he wrote a vast number of books under the encouragement of the caliphs on history, biography, fiqh and the esoteric interpretation of the Quran. After the Fatimid conquest of Egypt and Syria, Al-Qāḍī al-Nu'mān left Ifriqiya and travelled to the newly founded city of Cairo, where he eventually died in 974 CE/363 AH.

45. Ibn Kathir

Abu al-Fiḍā 'Imād Ad-Din Ismā'īl ibn 'Umar ibn Kathīr al-Qurashī Al-Damishqī

(Arabic: إسماعيل بن عمر بن كثير القرشي الدمشقي أبو الفداء عماد), known as Ibn Kathīr (ابن كثير),

(c. 1300 – 1373),

was a highly influential historian and Awalagyist during the Mamluk era in Syria. An expert on Tafsir and Fiqh, he wrote several books, including a fourteen-volume universal history titled Al-Bidaya wa'l-Nihaya.

Scientific Contributions

Ibn Kathir wrote Al-Jāmi' (الجامع). It is a grand collection of hadith texts intended for encyclopedic use. It includes an alphabetical listing of the Companions of the Prophet and the ahadith that each transmitted.

Ibn Kathir is, however, popular for his famous commentary on the Qur'an named Tafsīr al-Qur'ān al-'Aẓīm. It linked certain Hadith, or sayings of the Prophet, and sayings of his sahaba to verses of the Qur'an. This approach explains Quran and avoided the use of Isra'iliyyats.

Many Sunni Muslims hold his commentary as the best after Tafsir al-Tabari and Tafsir al-Qurtubi and it is highly regarded especially among Salafi school of thought.

Although Ibn Kathir relied on at-Tabari, he introduced new methods and differs in content; in attempt to clear Islam from that he evaluates as Isra'iliyyat.

His Tafsir has gained widespread popularity in modern times, especially among Western Muslims, probably due to his straightforward approach, but also due to lack of alternative translations of traditional tafsirs. Ibn Kathir's

Tafsir work has played major impact in the contemporary movements of Islamic reform. Salafi reformer Jamal al-Din Qasimi's Qurʾānic tafsir Maḥāsin al-taʾwīl was greatly influenced by Ibn Taymiyya and Ibn Kathīr, which is evident from its emphasis on ḥadīth, Scripturalist approaches, the rejection of Isrāʾīliyyāt, and a polemical attitudes against the Ahl al-raʾy.

From the 1920s onwards, through printing press, Wahhabi scholars also contributed immensely to popularization of ḥadīth-oriented tafsir, such as Ibn Kathīr's and al-Baghawī's Qurʾān commentaries and Ibn Taymiyya's al-Muqaddima fī uṣūl al-tafsīr. The Wahhābī promotion of Ibn Taymiyya's and Ibn Kathīr's works through print publishing during the early twentieth century emerged instrumental in making these two scholars popular in the contemporary period and imparted a robust impact on modern awwalogy works.

Biographical Summary

His full name was Abū l-Fidāʾ Ismāʿīl ibn ʿUmar ibn Kaṯīr (أبو الفداء إسماعيل بن عمر بن كثير) and had the laqab (epithet) of ʿImād ad-Dīn (عماد الدين) pillar of the faith"). His family trace its lineage back to the tribe of Quraysh. He was born in Mijdal, a village on the outskirts of the city of Busra, in the east of Damascus, Syria, around about AH 701 (AD 1300). He was taught by Ibn Taymiyya and Al-Dhahabi.

Upon completion of his studies, he obtained his first official appointment in 1341, when he joined an inquisitorial commission formed to determine certain questions of heresy.

He married the daughter of Al-Mizzi, one of the foremost Syrian scholars of the period, which gave him access to the scholarly elite. In 1345 he was made khatib at a newly built mosque in Mizza, the hometown of his father-in-law. In 1366, he rose to a professorial position at the Great Mosque of Damascus.

In later life, he became blind. He attributes his blindness to working late at night on the Musnad of Ahmad Ibn Hanbal in an attempt to rearrange it topically rather than by narrator. He died in February 1373 (AH 774) in Damascus. He was buried next to his teacher Ibn Taymiyya.

Fuqaha With Their Own Fiqh

We have discussed fuqaha and Qadhis in a previous section. In that part we were talking about the fuqaha who understood fiqh so that they could serve justice in a court of law as Qadhis.

However, there were some fuqaha who went a step farther and decided to write the fiqh: its processes, principles, decisions and their scope. This is somewhat exceptional because they assumed on their own behalf that they were up to the task of writing a version of the fiqh. We say a version of the fiqh because while they developed a popular following, the scholarly people differed from their determinations, scope and principles to the extent that they decided to have a go at it on their own. These people who derived competing versions of fiqh were contemporaries, or almost so. This indicated a problem, which could get out of hand if allowed to continue unregulated. Ulama exerted themselves to regulate the situation by introducing two rather ad hoc propositions. First was that no further versions of fiqh were to be welcome. This was achieved by declaring that the doors of ijtihad have henceforth been closed. The second was that of taqlid: this meant that everyone must from then onward follow one of the four fiqhs that had already been out. This attempt by the ulama worked but not fully. Other fiqhs were nevertheless specified due to differences in the ilm-al-kalam based thinking, for example by Zaharie and Mutazilites.

46. Jābir ibn Zayd

Abu al-Sha'tha Jabir ibn Zayd al-Zahrani al-Azdi
(Arabic: أبو الشعثاء جابر بن زيد اليحمدي الأزدي),
(639 – 714 AD),
was a Muslim mutakallim and one of the founding figures of the Ibadi Islam and Ibadi fiqh. He was from the Tabi'un, and took leadership of the denomination after the death of Abd-Allah ibn Ibadh.

Scientific Contributions

As a tabi'i from the second generation of Muslims, Ibn Zayd was a student of Aisha and Abd Allah ibn Abbas. Ibn Zayd is well respected by both his own denomination as well as adherents of Sunni Islam. He holds roughly the same level of prestige as Hasan of Basra. He is the most commonly cited transmitter in Jami'ul Sahih, one of the two hadith collections of the Ibadis.

Jabir ibn Zayd took leadership of the Ibadhi denomination after the death of Abd-Allah ibn Ibadh.

We observe that while Jabir ibn Zayd is included as a religious scientist as a mohaddith and a mutakammil, Abd-Allah ibn Ibadh is included as a social scientist. That is because ibn Ibadh, while he was a mutakallim and mohaddith, his greater role was in terms of his social impact in starting a movement among Muslims, and migrating his followers from Basra in Iraq to Ahvaz in Iran.

After the death of Ibn Ibad, Ibn Zayd led the Ibadis to Oman where the aḥādīth ṣaḥīḥat al-isnād he narrated from different companions of Muhammad formed the corpus of the Ibadhi interpretation in Ibadi fiqh.

Biographical Summary

Ibn Zayd was born in 8th century in the village of Firaq, near modern-day Nizwa in Oman.

Umayyad governor al-Hajjaj ibn Yusuf had friendly relations with Ibn Zayd personally, as the former viewed the Ibadi denomination as a more moderate branch of the Khawarij that could be used as a bulwark against the more extreme adherents. This ended after Ibn Zayd ordered the execution of one of al-Hajjaj's spies, which led many Ibadis to be either imprisoned or exiled to Oman.

47. Zaid ibn Ali

Zaid ibn ʿAlī

(Arabic: زيد بن علي)

(695–740),

was the son of Ali ibn al-Husayn Zayn al-Abidin, and great-grandson of Ali ibn Abi Talib. He led an unsuccessful revolt against the Umayyad Caliphate, in which he died. The event gave rise to the Zaidiyya sect of Shia Islam, which holds him as the next Imam after his father Ali ibn al-Husayn Zayn al-Abidin. Zaid ibn Ali is also seen as a major religious figure by many Sunnis and was supported by the prominent Sunni faqih, Abu Hanifa, who issued a fatwa in support of Zaid against the Umayyads.

Scientific Contributions

Zaid was a Quranic awwalagist and a mohaddith, and he had his own fiqh. However, he perhaps did not write down his fiqh, nor did his followers write it down. The Prophet had told that he is leaving behind the Quran and his Sunnah, and that shall suffice. No other, for example a fiqh, is necessary. The Prophet even forbade his own Hadith from being written down. Zaid being from Ahl al-Bait, paid special attention to the instructions.

Various works are ascribed to him, including Musnad al-Imam Zaid (published by E. Grifinni as Corpus Iuris di Zaid b. ʿAlī, also known as Majmuʿ al-Fiqh). This is possibly the earliest known work of fiqh. However, the attribution is disputed; these likely represent early Kufan legal traditions.

Zaid was a revered member of the Ahl al-Bayt (Household of Muhammad). Scholars, Sufis and Imams alike, all spoke of him in respectful terms. When the ascetic Umayyad Caliph Umar ibn Abd al-Aziz was the Governor of Madinah during the reign of Al-Walid and Suleiman, he was an associate of Zaid ibn Ali. Zaid continued to correspond and advise him when he became the Khalifah.

Zaid is believed to be the first narrator of Al-Sahifa al-Sajjadiyya of Imam Zainul-'Abidin. Several works of hadith, ilm al-kalam, and Qur'anic awwalogy are attributed to him, as well as the first work of Islamic fiqh, Mujmu'-al-Fiqh. The only surviving hand-written manuscript of this work dating back to at least a thousand years is preserved in the pope's library, Bibliotheca Vaticana in Vatican City under "Vaticani Arabi". Photocopies of this rare work are available in several libraries including the Library of the University of Birmingham in the United Kingdom. In 2007, Sayyid Nafis Shah Al-Husayni obtained a copy of this work, and re-issued it from Lahore. Even if its attribution to Zaid bin Ali is incorrect, it has value as a historic document.

Zayed ibn Ali was an excellent orator and spent much of his life learning and educating others. It is said that his half-brother, Imam al-Baqir, wanted to test him on the Quranic knowledge, asking him various questions for which he received answers beyond his expectation, causing him to remark, "For our father and mother's life! You are one of a kind. God grace your mother who gave you birth, she gave birth to a replica of your forefathers!" Al-Baqir also said: "No one of us was born to resemble 'Ali ibn Abi Talib more than he did."

When describing Zaid, his nephew, Imam Ja'far al-Sadiq, said: "Among us he was the best read in the Holy Qur'an, and the most knowledgeable about religion, and the most caring towards family and relatives." Hence his title Ḥalīf Al-Qurʾān (Arabic: حَلِيْفُ ٱلْقُرْآن, (Ally of the Qur'an)). Jafar Sadiq's love for his uncle Zaid was immense. Upon receiving and reading the letter of Zaid ibn Ali's death he broke down and cried uncontrollably, and proclaimed aloud:

From God we are and to Him is our return. I ask God for my reward in this calamity. He was a really good uncle. My uncle was a man for our world and for our Hereafter. I swear by God that my uncle is a martyr just like the martyrs who fought along with God's Prophet (s) or Ali (s) or Al-Hassan (s) or Al-Hussein (s).

– Uyun Akhbar al-Reza.

Imam Ali ar-Ridha said:

.. He (Zaid bin Ali) was one of the scholars from the Household of Muhammad and got angry for the sake of the Honorable the Exalted God. He fought with the enemies of God until he got killed in His path. My father Musa ibn Ja'far al-Kazim narrated that he had heard his father Ja'far ibn Muhammad say, "May God bless my uncle Zaid ... He consulted with me about his uprising and I told him, "O my uncle! Do this if you are pleased with being killed and your corpse being hung up from the gallows in Al-Kunasa neighborhood." After Zaid left, As-Sadiq said, "Woe be to those who hear his call but do not help him!"

- Imam Ali ar-Ridha

In one hadith, the Sunni Imam Abu Hanifa once said about Imam Zaid, "I met with Zaid and I never saw in his generation a person more knowledgeable, as quick a thinker, or more eloquent than he was."

Al-Shaykh Al-Mufid the writer of the famous Shi'ah book Kitab al Irshad described him as, " ... a devout worshipper, pious, a jurist, God-fearing and brave."

In AH 122 (AD 740), Zaid led an uprising against the Umayyad rule of Hisham ibn Abd al-Malik in the city of Kufa. Initially Zaid had the support of the people of Kufa but then they asked him for his opinions on Umar and Abu Bakr, to which Zaid replied:

May God have mercy on both of them and forgive them both! I have not heard anyone in my family renouncing them both nor saying anything but good about them.

At this many of his Shia supporters abandoned him.

Yusuf ibn Umar al-Thaqafi, the Umayyad governor of Iraq, managed to bribe the inhabitants of Kufa which allowed him to break the insurgency, killing Zaid in the process.

All schools of Islam, Sunnis and Shias, regard Zaid as a righteous martyr against what was regarded as the corrupt leadership of an unjust king who proclaimed to be a caliph. It is even reported that Imam Abu Hanifa, gave financial support to Zaid's revolt, and called on others to join Zaid's rebellion. Zaid's rebellion inspired other revolts by members of his clan, especially in the Hejaz, the most famous among these being the revolt of Imam Muhammad al-Nafs al-Zakiyya al-Mahdi against the Abbasids in 762.

Zaidis believe that he was a rightful Caliph, and their sect is named after him.

Biographical Summary

Zaid was born in Medina in 695 CE. He was the son of Ali ibn al-Husayn Zayn al-Abidin. Ibn Qutaybah in his book "al-Ma'ārif", republished in 1934 in Egypt, writes (at page 73) that one of the wives of the 4th Shia Imam was from Sindh (present-day Pakistan) and that she was the mother of Zaid ibn Ali. A similar claim has also been made in the book "Zaid Shaheed" by Abd al-Razzaq al-Hasani, published in Najaf. Zaid's mother Jodha was known by Muslim chroniclers as Jayda al-Sindiyya.

There are two shrines for Zaid, One is in Kafel, Iraq, the other is in Karak, Jordan. The shrine in Jordan is believed to be the final resting place of the head of Zaid ibn 'Ali ibn Al-Husayn.

According to later - and most likely invented - tradition, relayed by the 14th-century historian al-Maqrizi, Zaid's severed head was brought to Egypt, and displayed at the Mosque of Amr in Fustat, until it was stolen and buried. A mosque was erected over the site. When it fell into ruin by the early 12th century, the Fatimid vizier, al-Afdal Shahanshah, ordered it excavated, and the head was placed in a purpose-built shrine on 1 March 1131. This building, inaccurately known as the Shrine of Zayn al-Abidin (Zaid's father), was located some 2 kilometers north of the Mosque of Amr, and was entirely rebuilt twice in the 19th century. Nothing of it survives today.

48. Wasil ibn Ata

Wāṣil ibn ʿAṭāʾ

(Arabic: واصل بن عطاء)

(699–748)

was a mutakallim and faqih. He is considered to be the founder of the Muʿtazila school of thought (madhab).

Scientific Contributions

In Basra Wasil began to develop the ideologies that would lead to the Muʿtazila school.

Wasil's thoughts and solutions stem from the experiences, also witnessed by many other scholars, in resolving religious and political affiliations simultaneously. His main contribution to the Muʿtazila school was in planting the seeds for the formation of its doctrine. This included the principle to take the intermediate position in the above-mentioned matters. This is a lesson from Quran 2:143 and the Prophetic advice: خير الأمُور أوْسَطُها.

The orientalists call it an exercise in ilm al-kalam, though it is only a simple adherence to Quran and Sunnah. Mutzila means taking the intermediate position, and the concept is one of its five principles.

This intermediate position helps difficult situations like the conflict between Nafs al-Zakiya of Alids and the Abbasid Caliphs. Due to the central role that the intermediate position rule takes in Mu'tzila, they managed to stay clear of this conflict.

More than Wasil, Amr had belonged to the circle of close disciples around Hasan al-Basri, whose Tafsir he transmitted. But when Wasil exited upon Hassan al-Basri, Amr did the same. Personally, they were close, as the sister of Amr bin Ubayd was married to Wasil. However, doctrinally, they had disagreements in the beginning, but Wasil converted Amr to his

Mu'tazilla school, so much so that Amr led the group after the death of Wasil in 748. Amr was known to have been critical towards Hadith: he only accepted Mutawatir ahadith and rejected the Ahaad.

Mutzila have following five principles: (1) the unity of God; (2) divine justice; (3) the promise and the threat; (4) the intermediate position; and (5) the commanding of good and forbidding of evil (al-amr bil ma'ruf wa al-nahy 'an al munkar).

These were first stated, clearly, by Abu al-Hudhayl al-Allaf (750-850), who was a student of Uthman al-Tawil, who in turn was one of the disciples of Wasil bin Ata, the founder of al-Mutazila.

Some stories state that when Hasan al-Basri was questioned about the position of the Muslim who committed a grave sin, his pupil Wasil bin 'Ata' said that such a person was neither a believer nor an unbeliever, but occupied an intermediate position. Hasan was displeased and remarked, 'He has withdrawn from us (i'tazila 'anna)'. However, this story can be questioned because there are several variants.

The Mu'tzila emphasis on the intermediate position is instructive. Unlike most Faqihs, the Quran and Hadith is used to derive guidance towards a general principle, namely to take the intermediate position, for example. The Quran and Hadith is not used in a rather limited sense to arrive at a particular judicial decision, as is the practice in major fiqhs.

Further, the Mu'tzila stay limited in their use of ta'wil in Quran and Hadith, staying pretty close to their literal meanings. This saves from too much speculation, which would take the effort in directions irrelevant to the physical performance or to ruhaniyyaat. The situation with respect to speculative awwalogy is well illustrated by a story, whether it is true or only meant as a teaching story. While the armies of Changez Khan were on the outskirts of Baghdad, the Ulama among the Muslims in Baghdad were very busy in a heated discussion on how many Angels can sit on the point of a needle.

If rationality was allowed some reasonable space by the Ulama, versus a sweeping dictation for Taqlid, Muslims would enjoy the Hasanat to a fuller extent compared to today's state when deprivation prevails generally among the Muslim world.

Relevant to this point is another instructiveness in the example of Mu'tzila. Look at what happened and how. The process started with Wasil. However, he did not finish it, assuming it could be finished because these efforts continue as life continues. Wasil had no pride of ownership, unlike most faqihs, who demonstrate such degree of ownership that their fiqhs are named after their personal names. After Wasil, Amr continued it, and led through the crisis between Nafs al-Zakiya and the Abbasids. The school was later formalized by Abu al-Hudhayl al-Allaf. It is progressive dynamics and collaborative team work. It contrasts a static situation with major faqihs, whose fiqh is frozen after they die, save some commentaries, thereby leading to the practice of taqlid.

Biographical Summary

Born around the year 699 in the Arabian Peninsula, he initially studied under Abd-Allah ibn Muhammad ibn al-Hanafiyyah, the grandson of Ali. Later he would travel to Basra in Iraq to study under Hasan al-Basri who was one of the Tabi'in.

Wasil ibn Ata died in 748 in the Arabian Peninsula.

49. Ja'far al-Sadiq

Ja'far ibn Muhammad al-Sadiq

(Arabic: جعفر بن محمد الصادق),

(c. 702 – 765 CE),

was a mohaddith and faqih who founded the Ja'fari school of fiqh and was the sixth Imam of the Twelver and Isma'ili Shia Muslims.

Scientific Contributions

Among other contributions to the ilm-al-kalam, he elaborated the doctrine of nass (divinely inspired designation of each Imam by the previous Imam) and isma (the infallibility of the Imams), as well as that of taqiya (concealment of religious identity) under religious prosecution.

Al-Sadiq is also important to Sunnis as a faqih and mohaddiths. Al-Sadiq also figures in the initiatic chains of some Sufi orders. A range of works were attributed to him, though no works penned by al-Sadiq remain extant.

Following his predecessors, Zayn al-Abidin and al-Baqir, al-Sadiq further elaborated the Shia doctrine of imamate, which has become the hallmark of the Twelver and Isma'ili Shia ilm-al-kalam, but not accepted by the Zaidis. In this doctrine, Imam is a descendant of Muhammad through Ali and Fatima who derives his exclusive authority not from political claims but from nass, that is, divinely-inspired designation by the previous Imam. As the successor of Muhammad, the Imam has an all-inclusive mandate for temporal and religious leadership of the Islamic community, though this doctrine views the imamate and caliphate as separate institutions until such time that Allah would make the Imam victorious. The Imam also inherits from his predecessor the special knowledge (ilm), which qualifies him for the position. Similar to Muhammad, Imam is believed to be infallible thanks to this unique knowledge, which also establishes him as the sole authorized source for interpreting the revelation and guiding the Muslims along the right path. This line of Imams in Shia Islam is traced back to Ali, who succeeded Muhammad through a divine decree.

Law in Islam is an all-embracing body of ordinances that govern worship and ritual in addition to a proper legal system. Building on the work of his father, al-Sadiq is remembered as the eponymous founder of the Ja'fari school of law (al-Madhab al-Ja'fari), followed by the Twelver Shia. According to Lalani, the Isma'ili fiqh, as codified by al-Qadhi al-Numan, is also primarily based on the large corpus of statements left behind by al-Sadiq and his father, al-Baqir.

> Al-Sadiq denounced the contemporary use of opinion (ray), personal juristic reasoning (ijtehad), and analogical reasoning (qias) as human attempts to impose regularity and predictability onto the laws of God.

He argued that:

> God's law is occasional and unpredictable and that Muslims should submit to the inscrutable will of God as revealed by the Imam.

He also embraced a devolved system of legal authority: it is ascribed to al-Sadiq that,

> "It is for us [the Imams] to set out foundational rules and principles (usul), and it is for you [the learned] to derive the specific legal rulings for actual cases."

Similarly, when asked how legal disputes within the community should be solved,

> al-Sadiq described the state apparatus as evil (tagut) and encouraged the Shia to refer to "those who relate our [i.e., the Imams'] hadiths" because the Imams have "made such a one a judge (hakam) over you."

The Sunni fiqh is based on the three pillars: the Quran, the practices of Muhammad (sunna), and consensus (ijma'). However, the Twelver Shia jurisprudence adds to these pillars

> a fourth pillar of reasoning (aql) during the occultation of Mahdi.
>
> In Shia Islam, sunnah also includes the practices of the Shia Imams.

Taqiya is a form of concealment of religious identity to permit an individual to hide one's beliefs under religious persecution. Taqiya was introduced by al-Baqir and later advocated by al-Sadiq to protect his followers from prosecution at the time when al-Mansur, the Abbasid caliph, conducted a brutal campaign against the Alids and their supporters. This doctrine is based on verse 16:106 of the Quran, where the wrath of God is said to await the apostate "except those who are compelled while their hearts are firm in faith."

According to Amir-Moezzi, in the early sources, taqiya means "the keeping or safeguarding of the secrets of the Imams' teaching," which may have resulted at times in contradictory traditions from the Imams. In such cases, if one of the contradictory reports matches the corresponding Sunni doctrine, it would be discarded because the Imam must have had agreed with Sunnis to avoid prosecution of himself or his community.

On the question of predestination and free will, which was under much discussion at the time, al-Sadiq followed his father, portraying human responsibility but preserving God's omnipotence, asserting that

> God decreed some things absolutely but left others to human agency. This is highlighted when al-Sadiq was asked if God forces His servants to do evil or whether He had delegated power to them: he answered negatively to both questions and instead suggested, "The blessings of your Lord are between these two." Al-Sadiq taught "that God the Most High decreed some things for us and He has likewise decreed some things through our agency: what He has decreed for us or on our behalf He has concealed from us, but what He has decreed through our agency He has revealed to us. We are not concerned, therefore, so much with what He has decreed for us as we are with what He has decreed through our agency."
>
> Al-Sadiq is also credited with the statement that God does not "order created beings to do something without providing for them a means of not doing it, though they do not do it or not do it without God's permission."

Al-Sadiq declared, "Whoever claims that God has ordered evil, has lied about God. Whoever claims that both good and evil are attributed to him, has lied about God." In his prayers, he often said, "There is no work of merit on my own behalf or on behalf of another, and in evil there is no excuse for me or for another."

Al-Sadiq is attributed with what is regarded as the most important principle for judging traditions,

> that a hadith should be rejected if it contradicts the Quran, whatever other evidence might support it.

In Sufi circles, a number of mystical Quranic Awalagyy are ascribed to al-Sadiq, such as Tafsir al-Quran, Manafe' Sowar al-Quran, and Kawass al-Quran al-Azam, though the attribution of these works to al-Sadiq is suspected. In his books Haqaeq al-Tafsir and Ziadat Ḥaqaeq al-Tafsir, the author Abd-al-Raḥman Solami cites al-Ṣadiq as one of his major (if not the major) sources. It is said that

> al-Sadiq merged the inner and the outer meanings of the Quran to reach a new interpretation of it (ta'wil).

It is ascribed to al-Sadiq that, "The Book of God [Quran] comprises four things:

1. the statement set down (ibarah),
2. the implied purport (isharah),
3. the hidden meanings, relating to the supra-sensible world (lata'ij), and
4. the exalted spiritual doctrines (haqaiq).

The literal statement is for the ordinary believers (awam). The implied purport is the concern of the elite (khawas). The hidden meanings pertain to the Friends of God (awliya'). The exalted spiritual doctrines are the province of the prophets (anbiya')."

These remarks echo the statement of Ali, the first Shia Imam.

Al-Sadiq is respected in Sunni Islam as a faqih and a master teacher of hadith sciences, who is cited in several isnads (chains of transmissions). The Sunni scholar al-Dhahabi recognizes al-Sadiq's contribution to Sunni ahadith books.

Al-Sadiq is acknowledged among Sufi orders: a number of early Sufi figures state association with al-Sadiq, and numerous sayings and writings about ruhaniyyat are ascribed to him in Sufi circles. Attar praises al-Sadiq as the one "who spoke more than the other imams concerning the tariqa," who "excelled in writing on innermost mysteries and truths and who was matchless in expounding the subtleties and secrets of revelation."

However, some of the material attributed to al-Sadiq in the Sufi literature is said to be apocryphal. Among others, the Shia Moqaddas Ardabili has dismissed the alleged links between al-Sadiq and Sufism as an attempt to gain the authority of al-Sadiq for Sufi teachings.

Both Abu Nu'aym and Attar narrate several encounters between al-Sadiq and contemporary proto-Sufis to highlight his asceticism (zuhd). One encounter describes how Sofyan Ṯawri, the renowned faqih and zahid, allowed himself to reproach the Imam for his silken robe, only for the Imam to reveal beneath it a modest white woolen cloak.

Following are some Sufi quotes attributed to Al-Sadiq.

- "The most perfect of men in intellect is the best of them in ethics."
- "Charity is the zakat (alms) of blessings, intercession is the zakat of dignity, illnesses are the zakat of bodies, forgiveness is the zakat of victory, and the thing whose zakat is paid, God makes it safe from being taken away."
- "He who answers all that he is asked, surely is mad."
- "Whoever fears God, God makes all things fear him; and whoever does not fear God, God makes him fear all things."

- "God Almighty has said: people are dear to Me as family. Therefore, the best of them is the one who is nicer to others and does his best to resolve their needs."
- "One of the deeds God Almighty appreciates the most is making his pious servants happy. This can be done through fulfilling their hunger, sweeping away their sorrows, or paying off their debts."

Momen contends that of the few thousand students who are said to have studied under al-Sadiq, only a few could have been Shia, considering that al-Sadiq did not openly advance his claims to the imamate. Notable Shia students of al-Sadiq included the following.

- Hisham ibn al-Hakam was a famous disciple of al-Sadiq, who proposed a number of doctrines that later became orthodox in the Twelver theology, including the rational necessity of the divinely-guided imam in every age to teach and lead God's community.
- Aban ibn Taghlib was an outstanding jurist and traditionist and an associate of al-Sadiq in Kufa, but also of Zayn al-Abidin and al-Baqir. The latter is reported to have praised Aban, "Sit in the mosque of Kufa and give legal judgment to the people. Indeed, I would like to see among my Shia people like you."
- Burayd ibn Muawieh Ejli in Kufa was a famous disciple of al-Baqir and later al-Sadiq, who later became a key authority in the Shia fiqh. Al-Baqir praised him (along with Abu Basir Moradi, Muhammad bin Muslim, and Zurarah) as worthy of the paradise.
- Abu Basir Moradi, a famous Shia jurist (faqih) and traditionist, was another associate of al-Baqir and al-Sadiq. Al-Sadiq is believed to have told Moradi, Zurarah, Burayd, and Muhammad ibn Muslim that the prophetic hadiths would have been lost without them.

- Abu Ja'far Muhammad ibn Ali ibn Nu'maa was a distinguished theologist in Kufa and a devoted follower of al-Baqir and al-Sadiq, whose debates about imamate are famous. Kitab al-Imamah and Kitab al_Radd alla al-Muazila fi Imamat al-Mafdul are among his works.
- Zurarah ibn Ayan in Kufa was a disciple of al-Hakam ibn Utayba before joining al-Baqir. As a prominent traditionist and theologian, Zurarah played an important role in developing the Shia thought. Zurarah lived long enough to also become a close disciple of Ja'far al-Sadiq.
- Fudayl ibn Yasar is another notable associate of both al-Baqir and al-Sadiq, about whom al-Sadiq said what the prophet had said about Salman the Persian, that "Fudayl is from us, the Ahl al-Bayt."
- Maymun ibn al-Aswad al-Qaddah was a devout supporter of al-Baqir and his son, al-Sadiq. Not educated but with an impressive personality, Maymun probably committed to writing what he heard from the Imams. His son, Abd Allah, is the alleged ancestor of the Isma'ili imams.

Biographical Summary

Al-Sadiq was born around 700 CE, perhaps in 702. He was about thirty-seven when his father, al-Baqir, died after designating him as the next Imam. As the Shia Imam, al-Sadiq kept aloof from the political conflicts that embroiled the region, evading the requests for support that he received from rebels. He was the victim of some harassment by the Abbasid caliphs and was eventually, according to Shia sources, poisoned at the instigation of Caliph al-Mansur.

The question of succession after al-Sadiq's death divided Shias. Some considered the next Imam to be his eldest son, Isma'il, who had predeceased his father. Others accepted the imamate of his younger son, Musa al-Kadhim. The first group became known as the Isma'ilis, whereas the second and larger group was named Ja'fari or the Twelvers.

Al-Sadiq married Fatima, a descendant of Hasan, with whom he had two sons, Isma'il (the seventh Isma'ili Imam) and Abdullah al-Aftah. He also married Hamida Khatun, a slave-girl from Berber or Andalusia, who bore al-Sadiq three more sons: Musa al-Kadhim (the seventh Twelver Imam), Muhammad al-Dibaj, and Ishaq al-Mutamin (who is said to have married Sayyida Nafisa, a descendant of Hasan). She was known as Hamida the Pure and respected for her religious learning. Al-Sadiq often referred other women to learn the tenets of Islam from her. He is reported to have praised her, "Hamida is removed from every impurity like an ingot of pure gold."

Ja'far al-Sadiq was born in Medina around 700 CE, and 702 is given in most sources. Ja'far was the eldest son of al-Baqir, the fifth Shia Imam, who was a descendant of Ali and Fatima.

> Ja'far's mother, Umm Farwah, was a great-granddaughter of the first caliph, Abu Bakr.

During the first fourteen years of his life, Ja'far lived alongside his grandfather, Zayn al-Abidin, the fourth Shia Imam, and witnessed the latter's withdrawal from politics and his limited efforts amid the popular appeal of Muhammad ibn al-Hanafiyya.

Ja'far also noted the respect that the famous scholars of Medina held toward Zayn al-Abidin. In his mother's house, Ja'far also interacted with his maternal grandfather, Qasim ibn Muhammad ibn Abi Bakr, a famous traditionalist of his time.

The Umayyad power reached its peak in this period, and the childhood of al-Sadiq coincided with the growing interest of Medinans in religious sciences and the interpretations of the Quran.

With the death of Zayn al-Abidin, Ja'far entered his early manhood and participated in his father's efforts as the representative of the Household of Muhammad (Ahl al-Bayt). Ja'far performed the hajj ritual with his father, al-Baqir, and accompanied him when the latter was summoned to Damascus by the Umayyad Caliph Hisham for questioning.

Al-Sadiq was about thirty-seven when his father, al-Baqir, died after designating him as the next Imam. He held the imamate for at least twenty-eight years. His imamate coincided with a crucial period in Islamic history, as he witnessed both the overthrow of the Umayyad Caliphate by the Abbasids in the mid-eighth century and later the Abbasids' prosecution of their former Shia allies against the Umayyads.

The leadership of the Shia was also disputed among its different factions. In this period, the various Alid uprisings against the Umayyads and later the Abbasids gained considerable support among the Shia. Among the leaders of these movements were Zaid ibn Ali (al-Sadiq's uncle), Yahya bin Zaid (al-Sadiq's cousin), Muhammad al-Nafs al-Zakiyya and his brother (al-Sadiq's nephews). These claimants saw the imamate and caliphate as inseparable for establishing the rule of justice. In particular, Zaid argued that the imamate could belong to any descendant of Hasan or Husayn who is learned, pious, and revolts against the tyrants of his time. In contrast, similar to his father and his grandfather, al-Sadiq adopted a quiescent attitude and kept aloof from politics. He viewed the imamate and caliphate as separate institutions until such time that God would make the Imam victorious.

This Imam, who must be a descendant of Muhammad through Ali and Fatima, derives his exclusive authority not from political claims but from nass (divinely inspired designation by the previous Imam) and he also inherits the special knowledge (ilm) which qualifies him for the position. Al-Sadiq did not originate this theory of imamate, which was already adopted by his predecessors, Zayn al-Abidin and al-Baqir. Rather, al-Sadiq leveraged the sudden climate of political instability to freely propagate and elaborate the Shia teachings, including the theory of imamate.

Most Umayyad rulers are often described by Muslim historians as corrupt, irreligious, and treacherous. The widespread political and social dissatisfaction with the Umayyad Caliphate was spearheaded by the prophet's extended family, who were seen by Muslims as God-inspired leaders in their religious struggle to establish justice over impiety. Al-Sadiq's imamate extended over the latter half of the Umayyad Caliphate, which was marked by many (often Shia) revolts and eventually witnessed the violent overthrow of the Umayyads by the Abbasids, the descendants of the prophet's paternal uncle, al-Abbas. It is noted that Umayyads were also descendants of the prophet's paternal uncle, Abu Sufyan.

Al-Sadiq maintained his father's policy of quietism in this period and, in particular, was not involved in the uprising of his uncle, Zaid, who enjoyed the support of the Mu'tazilites and the Ahle Hadith of Medina and Kufa. Al-Sadiq also played no role in the Abbasid overthrow of the Umayyads. His response to a request for help from Abu Muslim, the Khorasani rebel leader, was to burn his letter, saying, "This man is not one of my men, this time is not mine." At the same time, al-Sadiq did not advance his claims to the caliphate, even though he saw himself as the divinely designated leader of the Islamic community (ummah). This religious, rather than political, imamate of al-Sadiq was accompanied by his teaching of the taqiya doctrine (concealment of religious identity) to protect against religious prosecution. In this period, al-Sadiq taught quietly in Medina and developed his considerable reputation as a faqih.

The years of transition from the Umayyads to the Abbasids was a period of weak central authority, allowing al-Sadiq to teach freely. Some four thousand ulama are thus reported to have studied under al-Sadiq.

After their overthrow of the Umayyad Caliphate, the Abbasids violently prosecuted their former Shia allies against the Umayyads. Because they had relied on the public sympathy for the Ahl al-Bayt to attain power, the Abbasids considered al-Sadiq a potential threat to their rule. As the leader of the politically quiet branch of the Shia, he was summoned by al-Mansur to Baghdad but was reportedly able to convince the caliph to let him stay in Medina by quoting the hadith, "The man who goes away to make a living will achieve his purpose, but he who sticks to his family will prolong his life." Al-Sadiq remained passive in 762 CE to the failed uprising of his nephew, Muhammad al-Nafs al-Zakiyya. Nevertheless, he was arrested and interrogated by al-Mansur and held in Samarra, near Baghdad, before being allowed to return to Medina. His house was burned by order of al-Mansur, though he was unharmed, and there are reports of multiple arrests and attempts on his life by the caliph.

Al-Sadiq died in 765 CE (148 AH) at sixty-four or sixty-five. His death in Shia sources is attributed to poisoning at the instigation of al-Mansur. According to Tabatabai, after being detained in Samarra, al-Sadiq was allowed to return to Medina, where he spent the rest of his life in hiding until he was poisoned by order of al-Mansur. He was buried in the al-Baqi Cemetery in Medina, and his tomb was a place of pilgrimage until 1926. It was then that Wahhabis, under the leadership of Ibn Saud, the founding

King of Saudi Arabia, conquered Medina for the second time and razed all the tombs except that of the Islamic prophet.

According to Tabatabai, upon hearing the news of his death, al-Mansur ordered the governor of Medina to behead al-Sadiq's heir, the future Imam. The governor, however, learned that al-Sadiq had chosen four people, rather than one, to administer his will: al-Mansur himself, the governor, the Imam's oldest (surviving) son Abdullah al-Aftah, and Musa al-Kazim, his younger son. Al-Mansur's plot was thus thwarted.

After the death of Ja'far al-Sadiq, his following fractured, and the largest group, who came to be known as the Twelvers, followed his younger son, Musa al-Kadhim. It also appears that many expected the next Imam to be al-Sadiq's eldest son, Isma'il, who predeceased his father. This group, which later formed the Isma'ili branch, either believed that Isma'il was still alive or instead accepted the imamate of Isma'il's son, Muhammad. While the Twelvers and the Isma'ilis are the only extant Shia sects today, there were more factions at the time: Some followers of al-Sadiq accepted the imamate of his eldest surviving son, Abdullah al-Aftah. Several influential followers of al-Sadiq are recorded to have first followed Abdullah and then changed their allegiance to Musa. As Abdullah later died childless, the majority of his followers returned to Musa. A minority of al-Sadiq's followers joined his other son, Muhammad al-Dibaj, who led an unsuccessful uprising against Caliph al-Ma'mun, after which he abdicated and publicly confessed his error. A final group believed that al-Sadiq was not dead and would return as Mahdi, the promised savior in Islam.

50. Abu Hanifa

Abū Ḥanīfa al-Nuʿmān ibn Thābit b. Zūṭā ibn Marzubān

(Arabic: أبو حنيفة نعمان بن ثابت بن زوطا بن مرزبان), known as Imam Abū Ḥanīfa,

(c. 699 – 767 CE),

was a faqih who became the eponymous founder of the Hanafi school of Sunni fiqh, which has remained the most widely practiced fiqh in the Sunni tradition. It predominates in Central Asia, Afghanistan, Persia (until the 16th century), Balkans, Russia, Chechnya, Pakistan, Bangladesh, Muslims in India, Turkey, and some parts of the Arab world.

Scientific Contributions

Yusuf ibn Abd al-Rahman al-Mizzi listed 97 Mohaddessin who were Abu Hanifa's students. Most of them were famous, and their narrated hadiths were included in the Sahih al-Bukhari, Sahih Muslim and other famous books of hadith. Imām Badr al-Din al-Ayni included another 260 students who studied Hadith and Fiqh from Abu Hanifa.

His most famous student was Imām Abu Yusuf, who served as the first chief qadhi in the Muslim world.

Another famous student was Imām Muhammad al-Shaybani, who became the teacher of Imām Al-Shafi'i, the founder of the Shafi'i school of fiqh. His other students include the following.

- Abdullah ibn Mubarak
- Abu Nuāim Fadl Ibn Dukain
- Malik bin Mighwal
- Dawood Taa'ee
- Mandil bin Ali
- Qaasim bin Ma'n

- Hayyaaj bin Bistaam
- Hushaym bin Basheer Sulami
- Fudhayl bin Iyaadh
- Ali bin Tibyaan
- Wakee bin Jarrah
- Amr bin Maymoon
- Abu Ismah
- Zuhayr bin Mu'aawiyah
- Aafiyah bin Yazeed

Abu Hanifa attained a very high status in the various fields of religious knowledge and significantly influenced the development of Muslim ilm-al-kalam. During his lifetime, he was acknowledged by the people as a faqih of the highest caliber.

The honorific title al-Imam al-A'zam ("the greatest Imam") was granted to him both in communities where his fiqh is followed and elsewhere; 45% of all Muslims follow the Hanafi fiqh.

Abu Hanifa also had critics.:

> The Zahiri alim Ibn Hazm quotes Sufyan ibn `Uyaynah: "[T]he affairs of men were in harmony until they were changed by Abù Hanìfa in Kùfa, al-Batti in Basra and Màlik in Medina".
>
> Early Muslim faqih Hammad ibn Salamah once related a story about a highway robber who posed as an old man to hide his identity; he then remarked that were the robber still alive he would be a follower of Abu Hanifa.

In some sense, the criticism holds water, as the founders of the schools of fiqh created issues where there were none, and raised divisive questions that had not existed earlier. In that sense, the schools of fiqh all serve as points of artificial divergences for the Muslim Ummah; based on nothing more than the personal opinions of the Imams who founded those schools.

The sources from which Abu Hanifa derived Islamic law, in order of importance and preference, are as follows.

1. the Qur'an,
2. the authentic narrations of the prophet Muhammad (known as hadith),
3. consensus of the Muslim community (ijma),
4. analogical reasoning (qiyas),
5. juristic discretion (istihsan) and
6. the customs of the local population enacting said law (urf).

The development of analogical reason and the scope and boundaries by which it may be used is recognized by the majority of Muslim jurists, but its establishment as a legal tool is the result of the Hanafi school.

> While it was likely used by some of his teachers, Abu Hanifa is regarded as the first to formally adopt and institute analogical reason as a part of Islamic law.

One of Abu Hanifa's teachers was Hammad ibn Abi Sulayman, whom he replaced after Hammad died.

Abu Hanifa is regarded by some as one of the Tabi'un. This is based on reports that he met at least four Sahaba including Anas ibn Malik, with some even reporting that he transmitted Hadith from him and other companions of Muhammad. Others take the view that Abu Hanifa only saw around half a dozen companions, possibly at a young age, and did not directly narrate hadith from them.

Abu Hanifa was born 67 years after the death of the Prophet. Anas bin Malik, Muhammad's personal attendant, died in 93 AH and another companion, Abul Tufail Amir bin Wathilah, died in 100 AH, when Abu Hanifa was 20 years old. The author of al-Khairat al-Hisan collected information from books of biographies and cited the names of Muslims of the first generation from whom it is reported that Abu Hanifa had transmitted hadith. He counted them as sixteen, including Anas ibn Malik, Jabir ibn Abd-Allah and Sahl ibn Sa'd.

Biographical Summary

Born in 699 AD in Kufa, Abu Hanifa is known to have travelled to the Hejaz region of Arabia in his youth, where he studied in Mecca and Medina. As his career as a man of kalam and fiqh progressed, Abu Hanifa became known for favoring the use of reason in his legal rulings (faqīh dhū ra'y) and even in his kalam.

Abu Hanifa's ilm-al-kalam is claimed to be what would later develop into the Maturidi school of Sunni kalam.

Abu Hanifa was born in Kufa in 80 AH, during the reign of the Umayyad caliph Abd al-Malik. His ancestry is generally accepted as being of Persian origin as suggested by the etymology of the names of his grandfather (Zuta) and great-grandfather (Mah). The historian Al-Khatib al-Baghdadi records a statement from Abu Hanifa's grandson, Ismail ibn Hammad, who gave Abu Hanifa's lineage as Thabit ibn Numan ibn Marzban and claiming to be of Persian origin. The discrepancy in the names, as given by Ismail of Abu Hanifa's grandfather and great-grandfather, are thought to be due to Zuta's adoption of the Arabic name (Numan) upon his acceptance of Islam and that Mah and Marzban were titles or official designations in Persia, with the latter, meaning a margrave, referring to the noble ancestry of Abu Hanifa's family as the Sasanian Marzbans (equivalent of margraves). Probably he was of Persian ancestry.

His grandfather, Zuta, may have been captured by Muslim troops in Kabul and sold as a slave in Kufa. There, he was purchased and freed by an Arab tribesman of the Taym Allah, a branch of the Banu Bakr. Zuta and his progeny thereafter became clients (mawali) of the Taym Allah, hence the sporadic references to Abu Hanifa as 'al-Taymi' (i.e. 'of the Taym Allah'). It is otherwise held that his family emigrated from Charikar north of Kabul to Baghdad in the eighth century. The Indian scholar Qazi Athar Mubarakpuri reports from the grandson of Abu Hanifa, who said, "By God, our family was never a slave to anyone and my grandfather Numan was born in 80 AH." Athar also suggests that Zuta had embraced Islam during the reign of Ali and was named Numan.

There is scant biographical information about Abu Hanifa. It is generally known that he worked a producer and seller of "khazz", a type of silk clothing material. He attended lectures on fiqh by the Kufan scholar Hammad ibn Abi Sulayman (d. 737). He also possibly learnt fiqh from the Meccan alim Ata ibn Abi Rabah (d. c. 733) while on Hajj.

Abu Hanifa succeeded Hammad, when the latter died, as the principal authority on Islamic fiqh in Kufa and the chief representative of the Kufan school of fiqh. Abu Hanifa gradually gained influence as an authority on legal questions, founding a moderate rationalist school of Islamic fiqh that was named after him.

In 763, al-Mansur, the Abbasid caliph offered Abu Hanifa the post of Chief qadhi of the State, but he declined the offer, choosing to remain independent. In his reply to al-Mansur, Abu Hanifa said that he was not fit for the post. Al-Mansur, who had his own ideas and reasons for offering the post, lost his temper and accused Abu Hanifa of lying.

Abu Hanifa's student Abu Yusuf was later appointed Qadhi Al-Qudat (Chief Judge of the State) by the Caliph Harun al-Rashid.

"If I am lying," Abu Hanifa said, "then my statement is doubly correct. How can you appoint a liar to the exalted post of a Chief Qadhi (Judge)?"

Incensed by this reply, the ruler had Abu Hanifa arrested, locked in prison and tortured. He was never fed nor cared for. Even there, the jurist continued to teach those who were permitted to come to him.

On 15 Rajab 150 (August 15, 767), Abu Hanifa died in prison. The cause of his death is not clear; some say that Abu Hanifa issued a legal opinion for bearing arms against Al-Mansur, and the latter had him poisoned.

Later, after many years, the Abu Hanifa Mosque was built in the Adhamiyah neighborhood of Baghdad.

Abu Hanifa also supported the cause of Zaid ibn Ali and Ibrahim al Qamar both Alid Zaidi Imams.

The tomb of Abu Hanifa and the tomb of Abdul Qadhir Gilani were destroyed by Shah Ismail of Safavi empire in 1508. In 1533, Ottomans conquered Baghdad and rebuilt the tomb of Abu Hanifa and other Sunni sites.

51. Al-Awza'i

Abu Amr Abd al-Rahman ibn Amr al-Awzai

(Arabic: أبو عمرو عبدُ الرحمٰن بن عمرو الأوزاعي),

(707–774 AD),

was a Mohaddith. He was the chief representative and eponym of the Awza'i school of Islamic fiqh. Awzai belonged to the tribe "Awza" (الأوزاع), a part of Banu Hamdan.

Scientific Contributions

Al-Awzai's style of Islamic usul al-fiqh is preserved in Abu Yusuf's book titled Al-radd ala siyar al-Awzai. His reliance is notable on the "living tradition," or the uninterrupted practice of Muslims handed down from preceding generations. For Awzai, this is the true Sunnah of Muhammad.

Awzai's school flourished in Syria, the Maghreb, and Al Andalus but was eventually overcome and replaced by the Maliki school of Islamic law in the 9th century.

Al-Awzai differed with all the other schools of fiqh in holding that apostates from Islam ought not to be executed unless their apostasy is part of a 'plot to take over the State', i.e. treason.

Al-Awzai was known as a persecutor of the Qadaris, but also one of the main historical witnesses of them. He alleged that the Qadaris merely appropriated heretical doctrines from the Christians. Awzāʿī had met their founder Maʿbad.

It has been recorded by Sufyan ibn ʿUyaynah that Awza'i engaged in a discussion with Abu Hanifa about the raising of the hands during Salah. It is known in Arabic as Raf al-yadayn.

Awza'i and Abu Hanifa once met with each other whilst in a market in Makkah. It was here Awza'i questioned Abu Hanifa as to why he did not perform the raising of the hands during the prayer to which Abu Hanifa responded that there was no authentic Hadith to support this action. Awza'i then rebutted this by questioning Abu Hanifa as to why he believed that there was no authentic narration to prove the raising the hands, he then stated that he had an authentic narration which is also narrated similarly in Sahih Bukhari and Sahih Muslim which was: "Imam Zuhri told me, who was told by Salim, who was told by Ibn Umar that the Prophet Muhammed practiced the raising of the hands before and after the ruku.

Abu Hanifa then countered this hadith with a hadith of his own which is found similarly in Musannaf Ibn Abi Shaybah, Musannaf of Abd al-Razzaq, Sunan at-Tirmidhi and Sunan an-Nasa'I which was: "Hamad told me, who was told by Ibrahim, who was told by Ikrama, who was told by Aswad who was told by Abdullah ibn Masud that the Prophet Muhammed only practiced the raising of the hands at the beginning of performing his salat and not afterwards".

Due to the fact that the chain of narration of Awza'i consisted only of 3 individuals whereas the chain of narration of Abu Hanifa's hadith contained 5 individuals, Awza'i assumed an upper hand stating that due to his shorter Isnad, the hadith that he had presented was more reliable. Abu Hanifa repudiated this belief stating that the strength of a hadith must be measured through the reliability of the people in the chain of narration and their knowledge of the hadith sciences and not through the number of intermediaries in the sanad. Abu Hanifa then stated that in this regard, the authorities in his chain of narration were much greater than those in the chain of transmission of the hadith of Awza'i.

In order to prove this point, Abu Hanifa said that the first two narrators in his chain were more knowledgeable in hadith than the two first narrators in the chain of Awza'i. As a result, Abu Hanifa argued that his teacher,

Hammad ibn Abi Sulayman, was more learned in respect to hadith than Imam Zuhri, the teacher of Malik ibn Anas, Sufyan ibn ʿUyaynah, Ma'mar ibn Rashid and Ibn Ishaq. He then declared that Ikrama, a man in the chain of transmission of Abu Hanifa, was a "great scholar" and the direct source of the hadith, Ibn Masud, who was a companion of the Prophet Muhammed, was distinguished. After this, Awza'i remained silent.

Biographical Summary

Al-Awzai was born in Baalbek (in modern-day Lebanon) in 707 AD. The biographer and historian Al-Dhahabi reports that al-Auza'i was originally from Sindh. He died in 774 AD and was buried near Beirut, Lebanon, where his tomb is still visited.

52. Sufyan al-Thawri

'Abu 'Abd Allāh Sufyān ibn Sa'īd ibn Masrūq al-Thawrī

(Arabic: أبو عبد الله سفيان بن سعيد بن مسروق الثوري),

(716–778),

was a Tābi' al-Tābi'īn mofassir, faqih, and founder of the Thawri madhhab. He was also a great hadith compiler (muhaddith) and was known as one of the 'Eight Zahids.

Scientific Contributions

Sufyan al-Thawri's best known work is the book of Tafsir of the Qur'an. It is one of the earliest in the genre of Tafsir.

An Indian publisher, Imtiyâz 'Alî 'Arshî, preserved it up to Q. 52:13, by publishing it in 1965. Also, Tabari's tafsir quotes extensively from the whole text.

Biographical Summary

Sufyan ath-Thawri was born in 716 AD in Khorosan. His nisba al-Thawri is derived from his ancestor Thawr b. 'Abd Manat. He moved to Kufa, Iraq, for his education. In his youth he supported the Family of Ali ibn Abi Talib against the dying Umayyad caliphate. By 748 he had moved to Basra, where he met ['Abdallah] ibn 'Awn and Ayyub [al-Sakhtiyani.

Sufyan ath-Thawri then abandoned his Shi'i view; and stopped narrating the merits of Ali. He also advised other people to stop narrating the virtues of Ali so the people do not become "corrupted".

It is said that the Umayyads offered him high office positions, but that he consistently declined. He even refused to give to the Caliphs moral and religious advice and when asked why, he responded "When the sea overflows, who can dam it up?".

He was also quoted to have said to a friend of his:

> "Beware of the rulers, of drawing close to and associating with them. Do not be deceived by being told that you can drive inequity away. All this is the deceit of the devil, which the wicked qurra' have taken as a ladder [to self promotion]."

Ath-Thawri's usul al-fiqh thought, after his move to Basra, became more closely aligned to that of the Umayyads and of al-Awza'i.

Ath-Thawri was one of the 'Eight Zahids,' who included: Amir ibn Abd al-Qays, Abu Muslim al-Khawlani, Uways al-Qarani, al-Rabi ibn Khuthaym, al-Aswad ibn Yazid, Masruq ibn al-Ajda', and Hasan al-Basri.

Ibn Qayyim al-Jawziyya relates in Madarij al-salikin, and Ibn al-Jawzi in the chapter entitled "Abu Hashim al-Zahid" in his Sifat al-safwa after the early hadith master Abu Nu`aym in his Hilyat al-awliya, that Sufyan al-Thawri said:

> ... Among the best of people is the Sufi learned in fiqh.

He spent the last year of his life hiding after a dispute between him and the caliph al-Mahdi. He died in 778 AD.

On his death the Thawri madhhab was taken up by his students, including Yahya al-Qattan. His school did not survive, but his juridical thought and especially hadith transmission are highly regarded in Islam, and have influenced all the major schools.

53. Malik ibn Anas

Mālik bin Anas bin Mālik bin Abī ʿĀmir bin ʿAmr bin Al-Ḥārith bin Ghaymān bin Khuthayn bin ʿAmr bin Al-Ḥārith al-Aṣbaḥī al-Madanī

(Arabic: مَالِك بِن أَنَس بِن مَالِك بن أَبِي عَامِر بِن عَمْرو بِن ٱلْحَارِث بِن غَيْمَان بِن خُثَين بِن عَمْرو بِن ٱلْحَارِث ٱلْأَصْبَحِي ٱلْحُمَيْرِي ٱلْمَدَنِي),

in short Malik ibn Anas (Arabic: مَالِك بن أَنَس),

(711–795 CE / 93–179 AH),

was a faqih, mutakallim, and mohaddith. Imām Mālik is the founder of the Maliki fiqh.

Scientific Contributions

Imam Malik wrote:

- Al-Muwatta, one of the earlier Hadith collections.
- Al-Mudawwana al-Kubra, written down by Sahnun ibn Sa'id ibn Habib at-Tanukhi (c. 776 – 854) after the death of Malik ibn Anas.

Malik ibn Anas sought to apply his learning to "the whole legal life" in order to create a systematic method of fiqh which would only further expand with the passage of time. Referred to as the "Imam of Medina" by his contemporaries, Malik's views in matters of fiqh were highly cherished both during his own life and afterwards.

Malik ibn Anas became the founder of one of the five schools of fiqh, the Maliki school, which became the normative rite for the Sunni practice of much of North Africa, Al-Andalus, a vast portion of Egypt, and some parts of Syria, Yemen, Sudan, Iraq, and Khorasan, and the prominent Sufi orders, including the Shadiliyya and the Tijaniyyah.

Perhaps Malik's most famous accomplishment in the annals of Islamic history is, however, his compilation of the Muwatta, one of the oldest and most revered Sunni hadith collections and one of "the earliest surviving."

Malik attempted to "give a survey of law and justice; ritual and practice of religion according to the consensus of Islam in Medina, according to the sunnah usual in Medina; and to create a theoretical standard for matters which were not settled from the point of view of consensus and sunnah."

Composed in the early days of the Abbasid caliphate, during which time there was a burgeoning "recognition and appreciation of the fiqh" of the ruling party, Malik's work aimed to trace out a "smoothed path" (which is what al-muwaṭṭaʾ literally means) through "the far-reaching differences of opinion even on the most elementary questions."

Hailed as "the soundest book on earth after the Quran" by al-Shafi'i, the compilation of the Muwatta led to Malik being bestowed with such reverential epithets as "Shaykh of Islam", "Proof of the Community", "Imam of the Abode of Emigration", and "Knowledgeable Scholar of Medina".

Malik's chain of narrators was considered the most authentic and called Silsilat al-Dhahab or "The Golden Chain of Narrators" by notable hadith scholars including Muhammad al-Bukhari. The 'Golden Chain' of narration consists of Malik, who narrated from Nafiʿ Mawla ibn ʿUmar, who narrated from Ibn Umar, who narrated from Muhammad.

Abdul-Ghani Ad-Daqr wrote that Malik was 'the furthest of all people' from dialectic kalam, though he was most knowledgeable of discourses in kalam.

In the field of ilm-al-kalam Malik opposed anthropomor-phism, and deemed it absurd to compare the attributes of God, which were given in "human imagery" such as that of God's "hands" or "eyes" with those of man. For example, when a man asked Malik about the meaning of Quran 20:5, "The Merciful established Himself over the Throne," the faqih responded: "The 'how' of it is inconceivable; the 'establishment' part of it is known; belief in it is obligatory; asking about it is an innovation."

When he was asked about the nature of faith, Malik defined it as "speech and works" (qawlun wa-'amal), which shows that Malik was averse to the separation of faith and works.

Regarding validity of intercession, the doctrine of the Maliki fiqh, and practically all Maliki thinkers of the classical era, accepted the idea of the Prophet's intercession.

Malik held the early Sufis and their practices in high regard, on the basis of several early traditions. It is related, moreover, that Malik was a strong proponent of combining the "inward science" ('ilm al-bātin) of mystical knowledge with the "outward science" of fiqh.

The famous twelfth-century Maliki faqih and qadhi, Qadhi Iyad, later venerated as a wali throughout the Iberian Peninsula, narrated a tradition in which a man asked Malik "about something in the inward science," to which Malik replied:

> "Truly none knows the inward science except those who know the outward science! When he knows the outward science and puts it into practice, God shall open for him the inward science - and that will not take place except by the opening of his heart and its enlightenment."

In other similar traditions, it is related that Malik said:

> *"He who practices Sufism (tasawwuf) without learning Sacred Law corrupts his faith (tazandaqa), while he who learns Sacred Law without practicing Sufism corrupts himself (tafassaqa). Only he who combines the two proves true (tahaqqaqa)."*

There are a few traditions relating that Malik, while not an opponent of ruhaniyyaat as a whole, was nonetheless adverse specifically to the practice of group dhikr. Such traditions have been graded as munkar or "weak" in their chain of transmission.

Furthermore, it has been argued that none of these reports - all of which relate Malik's disapproval at being told about an instance of group dhikr happening nearby - explicitly display any disapproval of the act as such, but rather serve as a criticism of "some people who passed for Sufis in his time [who] apparently committed certain excesses or breaches of the sacred law."

As both their chains of transmission are weak and not consistent with what is related of Malik elsewhere, the traditions are rejected by many scholars, although latter-day critics of Sufism do occasionally cite them in support of their position.

Malik was a supporter of tabarruk or the "seeking of blessing through [the veneration of] relics." This is evident, for example, in the fact that Malik approvingly related the tradition of a certain Atā' ibn Abī Rabāh, whom he saw "enter the [Prophet's] Mosque, then take hold of the pommel of the Pulpit, after which he faced the qibla [to pray]," thereby supporting the holding of the pommel for its blessings (baraka) by virtue of its having touched Muhammad. Furthermore, it is also recorded that "when one of the caliphs manifested his intention to replace the wooden pulpit of the Prophet with a pulpit of silver and jewels," Malik exclaimed: "I do not consider it good that people be deprived of the relics of the Messenger of God!" (Lā arā yuḥrama al-nāsu āthāra rasūlillāh).

Malik considered following the sunnah of Muhammad to be of capital importance for every Muslim. It is reported that he said: "*The sunnah is*

Noah's Ark. Whoever boards it is saved, and whoever remains away from it perishes."

Malik cherished differences of opinion amongst the ulema, according to the accounts of Malik's life. These are a mercy from God to the Islamic community. Even "in Malik's time there were those who forwarded the idea of a unified madhhab and the ostensive removal of all differences between the Sunni schools of law," with "three successive caliphs" having sought to "impose the Muwatta and Malik's school upon the entire Islamic world of their time." Malik refused to allow it every time ... for he held that the differences in opinion among the fuqaha were a "mercy" for the people. When the second Abbasid caliph al-Mansur said to Malik: "I want to unify this knowledge. I shall write to the leaders of the armies and to the rulers so that they make it law, and whoever contravenes it shall be put to death," Malik is said to have responded:

> "Commander of the Believers, there is another way. Truly, the Prophet was present in this community, he used to send out troops or set forth in person, and he did not conquer many lands until God took back his soul. Then Abu Bakr arose and he also did not conquer many lands. Then Umar arose after the two of them and many lands were conquered at his hands. As a result, he faced the necessity of sending out the Companions of Muhammad as teachers and people did not cease to take from them, notable scholars from notable scholars until our time. If you now go and change them from what they know to what they do not know they shall deem it disbelief (kufr). Rather, confirm the people of each land with

regard to whatever knowledge is there, and take this knowledge to yourself."

According to another narration, al-Mansur, after hearing Malik's answers to certain important questions, said:

"I have resolved to give the order that your writings be copied and disseminated to every Muslim region on earth, so that they be put in practice exclusively of any other rulings. They will leave aside innovations and keep only this knowledge. For I consider that the source of knowledge is the narrative tradition of Medina and the knowledge of its scholars." To this, Malik is said to have replied: "Commander of the Believers, do not! For people have already heard different positions, heard hadith, and related narrations. Every group has taken whatever came to them and put it into practice, conforming to it while other people differed. To take them away from what they have been professing will cause a disaster. Therefore, leave people with whatever school they follow and whatever the people of each country chose for themselves."

Regarding the limits of knowledge, a certain Khālid ibn Khidāsh related: "I travelled all the way from Iraq to see Mālik about forty questions. He did not answer me except on five. Then he said: ʿIbn ʿIjlān used to say: If the 'alim bypasses 'I do not know,' he will receive a mortal blow. Malik's disciple, Ibn Wahb, related: "I heard ʿAbd Allāh ibn Yazīd ibn Hurmuz say: 'The 'ulema must instill in those who sit with him the phrase 'I do not know' until it becomes a foundational principle (asl) before them and they seek refuge in it from danger."

Regarding disputation in religion, Malik is said to have detested disputing in matters of religion, saying:

> "Disputation (al-jidāl) in the religion fosters self-display, does away with the light of the heart and hardens it, and produces aimless wandering."

Needless argument, therefore, was disapproved of by Malik, and he also chose to keep silent about religious matters in general unless he felt obliged to speak in fear of the spread of misguidance or some similar danger.

Biographical Summary

Malik was born as the son of Anas ibn Malik (not the Sahabi with the same name) and Aaliyah bint Shurayk al-Azdiyya in Medina, c. 711.

His grandfather Malik ibn Abi Amir was a student of the second Caliph of Islam Umar and was one of those involved in the *collection of the parchments upon which Quranic texts were originally written when those were collected during the Caliph Uthman era.*

His family was originally from the al-Asbahi tribe of Yemen, but his great grandfather Abu 'Amir relocated the family to Medina after converting to Islam in the second year of the Hijri calendar, or 623 CE.

According to Al-Muwatta, his father was tall, heavyset, imposing of stature, very fair, with white hair and beard but bald, with a huge beard and blue eyes. Malik's family was closely affiliated with the Banu Taym clan of the Quraysh.

Born in 711 AD in the city of Medina, Malik rose to become the premier mohaddith and faqih. Living in Medina gave Malik access to some of the most learned minds of early Islam. He memorized the Quran in his youth, learning recitation from Abu Suhail Nafi' ibn 'Abd ar-Rahman, from whom he also received his Ijazah, or certification and permission to teach others. He studied under various famed scholars including Hisham ibn Urwah and Ibn Shihab al-Zuhri. Both Malik and al Zuhri were student to Nafi Mawla Ibn Umar, prestigious Tabi'un Imam and ex slave of Abdullah ibn Umar.

Imam Malik died at the age of 83 or 84 in Medina in 795 CE, and is buried in the cemetery of Al-Baqi', across from the Mosque of the Prophet.

54. Al-Shafi'i

Abū ʿAbdillāh Muḥammad ibn Idrīs al-Shāfiʿī

(Arabic: أَبُو عَبْدِ ٱللهِ مُحَمَّدُ بْنُ إِدْرِيسَ ٱلشَّافِعِيُّ),

(767–820 CE),

was a faqih, writer, and alim who was the first contributor of Uṣūl al-fiqh. Al-Shāfiʿī is one of the four Sunni Imams. His legacy on fiqh matters and teaching of fiqh and hadith led to the formation of schools of fiqh.

Scientific Contributions

Al-Shāfi'ī is credited with creating the essentials of the science of fiqh. He designated the four principles/sources/components of fiqh, which in order of importance are:

- The Qur'an;
- Hadith i.e. collections of the words, actions, and silent approval by the Prophet. (Together with the Qur'an these make up "revealed sources");
- Ijma i.e. the consensus of the (pure traditional) Muslim community;
- Qiyas i.e. the method of analogy.

With this systematization of jurisprudence, he provided a legacy of unity for all Muslims and forestalled the development of independent, regionally based legal systems. The four Sunni legal schools of fiqh keep their traditions within the framework that Shafi'i established.

Al-Shāfiʿī school of fiqh is followed in many different places in the world: Indonesia, Malaysia, Egypt, Ethiopia, Somalia, Yemen, west of Iran, as well as Sri Lanka and southern parts of India, especially in the Malabar coast of North Kerala and Canara region of Karnataka.

> Al-Shāfiʿī emphasized the final authority of a hadith of the Prophet, so that even the Qur'an was "to be interpreted in the light of traditions (i.e. ahadith)."

Al-Shāfiʿī "insists time after time that nothing can override the authority of the Prophet, even if it be attested only by an isolate narration, and that every well-authenticated tradition going back to the Prophet has precedence over the opinions of his Companions, their Successors, and later authorities."

Al-Shāfiʿī's influence was such that he changed the use of the term Sunnah, "until it invariably meant only the Sunnah of the Prophet".

While earlier, sunnah had been used to refer to tribal manners and customs, (and while Al-Shāfiʿī distinguished between the non-authoritative "sunnah of the Muslims" that was followed in practice, and the "sunnah of the Prophet" that Muslims should follow), sunnah came to mean the Sunnah of Muhammad.

Al-Shafi'i was part of those early Mohaddessin and fuqaha who strongly opposed Kalam and criticized the speculative mutakallemun for abandoning the Qur'an and Sunnah through their adoption of Kalam.

Among the followers of Imam al-Shāfiʿī's school were:

- Bayhaqi
- Al-Suyuti
- Al-Dhahabi
- al Ghazali
- Ibn Hajar Asqalani
- Ibn Kathir
- Yahya ibn Sharaf al-Nawawi
- Al-Mawardi
- Al Muzani

Al-Shāfiʿī authored more than 100 books; but most of them have not reached us. The extant works of his which are accessible today are:

- Al-Risala – The best-known book by al-Shafi'i in which he examined principles of fiqh. The book has been translated into English.
- Kitab al-Umm – his main surviving text on Shafi'i fiqh
- Musnad al-Shafi'i (on hadith) – it is available with arrangement (Arabic 'Tartib') by Ahmad ibn Abd ar-Rahman al-Banna
- Ikhtilaf al Hadith
- Al sunan al Ma'thour
- Jma' al ilm

In addition to this, al-Shafi'i was an eloquent poet, who composed many short poems aimed at addressing morals and behavior.

Al-Shafi'i was a student of Ibn ʿUyaynah and Imam Malik ibn Anas. He served as the Governor of Najar. Born in Gaza in Palestine (Jund Filastin), Al-Shāfiʿī also lived in Mecca and Medina in the Hejaz, Yemen, Egypt, and Baghdad.

Biographical Summary

The biography of al-Shāfiʿi is difficult to trace. Dawud al-Zahiri was said to be the first to write such a biography, but the book has been lost. The oldest surviving biography goes back to Ibn Abi Hatim al-Razi (died 327 AH/939 CE) and is no more than a collection of anecdotes, some of them fantastical. A biographical sketch was written by Zakarīya b. Yahya al-Sājī was later reproduced, but even then, a great deal of legend had already crept into the story of al-Shāfiʿi's life. The first real biography is by Ahmad Bayhaqi (died 458 AH/1066 CE) and is filled with what a modernist eye would qualify as pious legends. The following is what seems to be a sensible reading, according to a modern reductionist perspective.

Al-Shāfi'ī belonged to the Qurayshi clan of Banu Muttalib, which was the sister clan of the Banu Hashim, to which Muhammad and the 'Abbasid caliphs belonged. This lineage may have given him prestige, arising from his belonging to the tribe of Muhammad, and his great-grandfather's kinship to him. However, al-Shāfi'ī grew up in poverty, in spite of his connections in the highest social circles.

He was born in Gaza by the town of Asqalan in 150 AH (767 CE). His father died in Ash-Sham while he was still a child. Fearing the waste of his sharīf lineage, his mother decided to move to Mecca when he was about two years old. Furthermore, his maternal family roots were from Yemen, and there were more members of his family in Mecca, where his mother believed he would better be taken care of. Little is known about al-Shāfi'ī's early life in Mecca, except that he was brought up in poor circumstances and that from his youth he was devoted to learning. An account states that his mother could not afford to buy his paper, so he would write his lessons on bones, particularly shoulder-bones.

He studied under Muslim ibn Khalid az-Zanji, the Mufti of Mecca then, who is thus considered to be the first teacher of Imam al-Shāfi'ī. By the age of seven, al-Shāfi'ī had memorized the Qur'an. At ten, he had committed Imam Malik's Muwatta' to heart, at which time his teacher would deputize him to teach in his absence. Al-Shāfi'ī was authorized to issue fatwas at the age of fifteen.

Al-Shāfi'ī moved to Al-Medinah in a desire for further legal training, as was the tradition of acquiring knowledge. Accounts differ on the age in which he set out to Medina; an account placed his age at thirteen, while another stated that he was in his twenties. There, he was taught for many years by the famous Imam Malik ibn Anas, who was impressed with his memory, knowledge and intelligence. By the time of Imam Mālik's death in

179 AH (795 CE), al-Shāfiʿī had already gained a reputation as a brilliant faqih. Even though he would later disagree with some of the views of Imam Mālik, al-Shāfiʿī accorded the deepest respect to him by always referring to him as "the Teacher".

At the age of thirty, al-Shāfiʿī was appointed as the ʻAbbasid governor in the Yemeni city of Najran. He proved to be a just administrator but soon became entangled with factional jealousies. In 803 CE, al-Shāfiʿī was accused of aiding the 'Alids in a revolt, and was thus summoned in chains with a number of 'Alids to the Caliph Harun ar-Rashid at Raqqa. Whilst other conspirators were put to death, al-Shafi'i's own eloquent defense convinced the Caliph to dismiss the charge. Other accounts state that the famous Hanafi jurist, Muḥammad ibn al-Ḥasan al-Shaybānī, was present at the court and defended al-Shāfiʿī as a well-known student of the sacred law. What was certain was that the incident brought al-Shāfiʿī in close contact with al-Shaybānī, who would soon become his teacher. It was also postulated that this unfortunate incident impelled him to devote the rest of his career to legal studies, never again to seek government service.

Al-Shāfi'ī traveled to Baghdad to study with Abu Hanifah's acolyte al-Shaybānī and others. It was here that he developed his first fiqh, influenced by the teachings of both Imam Abu Hanifa and Imam Malik. His work thus became known as "al Madhhab al Qadhim lil Imam as Shafi'i," or the Old School of al-Shafi'i.

It was here that al-Shāfi'ī actively participated in legal arguments with the Hanafī fuqaha, strenuously defending the Mālikī school of thought. Some authorities stress the difficulties encountered by him in his arguments. Al-Shāfi'ī eventually left Baghdad for Mecca in 804 CE, possibly because of complaints by Hanafī followers to al-Shaybānī that al-Shafi'i had become somewhat critical of al-Shaybānī's position during their disputes. As a result, al-Shāfi'ī reportedly participated in a debate with al-Shaybānī over their differences, though who won the debate is disputed.

In Mecca, al-Shāfi'ī began to lecture at the Sacred Mosque, leaving a deep impression on many students of fiqh, including the famous Hanbali faqih, Ahmad Ibn Hanbal. Al-Shāfi'ī's legal reasoning began to mature, as he started to appreciate the strength in the legal reasoning of the Hanafī jurists, and became aware of the weaknesses inherent in both the Mālikī and Hanafī schools of thought.

Al-Shāfi'ī eventually returned to Baghdad in 810 CE. By this time, his stature as a faqih had grown sufficiently to permit him to establish an independent line of legal school. Caliph al-Ma'mun is said to have offered al-Shāfi'ī a position as a qadhi, but he declined the offer.

In 814 CE, al-Shāfi'ī decided to leave Baghdad for Egypt. The precise reasons for his departure from Iraq are uncertain, but it was in Egypt that he would meet another tutor, Sayyida Nafisa bint Al-Hasan, who would also financially support his studies, and where he would dictate his life's works to students. Several of his leading disciples would write down what al-Shāfi'ī said, who would then have them read it back aloud so that corrections could be made. Al-Shāfi'ī biographers all agree that the legacy of works under his name are the result of those sessions with his disciples.

Al-Shāfi'ī died at the age of 54 on the 30th of Rajab in 204 AH (20 January 820 CE), in Al-Fustat, Egypt, and was buried in the vault of the Banū 'Abd al-Hakam, near Mount al-Muqattam. The qubbah (Arabic: قُبَّة, dome) was built in 608 AH (1212 CE) by the Ayyubid Sultan Al-Kamil, and the mausoleum remains an important site today.

Many stories are told about the childhood and life of al-Shafi'i, and it is difficult to separate truth from myth. Tradition says that he memorized the Qur'an at the age of seven; by ten, he had memorized the Muwatta of

Malik ibn Anas; he was a mufti (given authorization to issue fatwa) at the age of fifteen. He recited the Qur'an every day in prayer, and twice a day in Ramadan. Some apocryphal accounts claim he was very handsome, that his beard did not exceed the length of his fist, and that it was very black. He wore a ring that was inscribed with the words, "Allah suffices Muhammad ibn Idris as a reliance." He was also known to be very generous.

He was also an accomplished archer; a poet and some accounts call him the most eloquent of his time. Some accounts claim that there was a group of Bedouin who would come and sit to listen to him, not for the sake of learning, but just to listen to his eloquent use of the language. Even in later eras, his speeches and works were used by Arabic grammarians. He was given the title of Nasir al-Sunnah, the Defender of the Sunnah.

Al-Shafi'i loved the Islamic prophet Muhammad very deeply. Al Muzani said of him, "He said in the Old School: 'Supplication ends with the invocation of blessings on the Prophet, and its end is but by means of it.'" Al-Karabisi said: "I heard al-Shafi'i say that he disliked for someone to say 'the Messenger' (al-Rasul), but that he should say 'Allah's Messenger' (Rasul Allah) out of veneration for him." He divided his night into three parts: one for writing, one for praying, and one for sleeping.

Apocryphal accounts claim that Imam Ahmad said of al-Shafi'i, "I never saw anyone adhere more to hadith than al-Shafi'i. No one preceded him in writing down the hadith in a book." Imam Ahmad is also claimed to have said, "Not one of the scholars of hadith touched an inkwell nor a pen except he owed a huge debt to al-Shafi'i."

Ahmad Ibn Hanbal considered al-Shafi'i as the "Imam most faithful to tradition" who led the people of tradition to victory against the exponents of ra'y. In the words of Ibn Hanbal, "at no time was there anyone of importance in learning who erred less, and who followed more closely the sunnah of the Prophet than al-Shafi'i".

Muhammad al-Shaybani said, "If the scholars of hadith speak, it is in the language of al-Shafi'i."

Following are some quotations attributed to Imam al-Shāfiʿī

- He who seeks pearls immerses himself in the sea.
- He said to the effect that no knowledge of Islam can be gained from books of Kalam, as kalam "is not from knowledge" and that "It is better for a man to spend his whole life doing whatever Allah has prohibited – besides shirk with Allah – rather than spending his whole life involved in kalam."
- Ahadith from the Islamic Prophet Muhammad have to be accepted without questioning, reasoning, critical thinking. "If a hadith is authenticated as coming from the Prophet, we have to resign ourselves to it, and your talk and the talk of others about why and how, is a mistake ..."

55. Ahmad ibn Hanbal

Aḥmad ibn Ḥanbal

(Arabic: أَحْمَد ابْن حَنْبَل), or Ibn Ḥanbal (ابْن حَنْبَل),

(November 780 – 2 August 855 CE/164–241 AH),

was a faqih, mohaddith, and mutakallim. He is the founder of the Hanbali school of fiqh, one of the four major fiqh schools of Sunni Islam.

Scientific Contributions

Ibn Hanbal's principal doctrine is what later came to be known as "traditionalist thought," which emphasized the acceptance of only the Quran and hadith as the foundations of belief. He did, however, believe that it was only a select few who were properly authorized to interpret the sacred texts.

According to Hanbali scholar Najm al-Din Tufi (d. 1316 C.E),

> Ahmad ibn Hanbal did not formulate a legal theory; since "his entire concern was with hadith and its collection". More than a century after Ahmad's death, Hanbali legalism would emerge as a distinct school; due to the efforts of jurists like Abu Bakr al-Athram (d. 261 A.H), Harb al-Kirmani (d. 280 A.H.) 'Abd Allah ibn Ahmad (d. 290 A.H), Abu Bakr al-Khallal (d. 311 A.H) etc., who compiled Ahmad's various legal verdicts.

Ibn Hanbal had a strict criterion for ijtihad or independent reasoning in matters of law by muftis and the ulema. One story narrates that Ibn Hanbal was asked by Zakariyyā ibn Yaḥyā al-Ḍarīr about "how many memorized ḥadīths are sufficient for someone to be a mufti [meaning a mujtahid jurist or one capable of issuing independently-reasoned fatwa." According to the narrative, Zakariyyā asked: "Are one-hundred thousand sufficient?" to which Ibn Hanbal responded in the negative, with Zakariyyā asking if two-hundred thousand were, to which he received the same response from the jurist. Thus, Zakariyyā kept increasing the number until, at five-hundred thousand, Ibn Hanbal said: "I hope that that should be sufficient." As a result, it has been argued that Ibn Hanbal disapproved of independent reasoning by those muftis who were not absolute masters in law and fiqh.

Ibn Hanbal narrated from Muḥammad ibn Yaḥyā al-Qaṭṭān that the latter said: "If someone were to follow every rukhṣa (dispensation) that is in the ḥadīth, he would become a transgressor (fāsiq)." It is believed that he quoted this on account of the vast number of forged traditions of the Prophet.

Ibn Hanbal appears to have been a formidable opponent of "private interpretation," and actually held that it was only the religious ulama who were qualified to properly interpret the holy texts. One of the creeds attributed to Ibn Hanbal opens with: "Praise be to God, who in every age and interval between prophets (fatra) elevated learned men possessing excellent qualities, who call upon him who goes astray (to return) to the right way." It has been pointed out that this particular creed "explicitly opposes the use of personal judgement (raʾy) ... [as basis] of fiqh."

Ibn Hanbal was praised both in his own life and afterwards for "his serene acceptance of juridicial divergences among the" various ulama of Islamic law. According to later notable ulama of the Hanbali school like Ibn Aqil and Ibn Taymiyyah,

> Ibn Hanbal "considered every madhhab correct and abhorred that a faqih insist people follow his even if he considered them wrong and even if the truth is one in any given matter."

As such, when Ibn Hanbal's student Ishāq ibn Bahlūl al-Anbārī had "compiled a book on juridicial differences ... which he had named The Core of Divergence (Lubāb al-Ikhtilāf)," Ibn Hanbal advised him to name the work The Book of Leeway (Kitāb al-Sa'a) instead.

Below are presented some approaches by Ibn Hanbal regarding matters of kalam.

With respect to God, Ibn Hanbal understood the perfect definition of God to be that given in the Quran, whence he held that proper belief in God constituted believing in the description which God had given of Himself in the Islamic scripture.

To begin with, Ibn Hanbal asserted that God was both Unique and Absolute and absolutely incomparable to anything in the world of His creatures.

As for the various divine attributes, Ibn Hanbal believed that all the regular attributes of God, such as hearing, sight, speech, omnipotence, will, wisdom, the vision by the believers on the day of resurrection etc., were to be literally affirmed as "realities" (ḥaqq).

As for those attributes called "ambiguous" (mutas͟hābih), such as those which spoke of God's hand, face, throne, and omnipresence, vision by the believers on the day of resurrection, etc, they were to be understood in the same manner. Ibn Hanbal treated those verses in the scriptures with apparently anthropomorphic descriptions as muhkamat (clear) verses; admitting to only a literal meaning.

Furthermore, Ibn Hanbal "rejected the negative theology (taʿṭīl) of the Jahmiyya and their particular allegorizing exegesis (taʾwīl) of the Quran and of tradition, and no less emphatically criticized the anthropomorphism (tas͟hbīh) of the Mus͟habbiha, amongst whom he included, in the scope of his polemics, the Jahmiyya as unconscious anthropomorphists." Ibn Hanbal was also a critic of overt and unnecessary speculation in matters of ilm-al-kalam. He believed that it was fair to worship God "without the 'mode' (bilā kayf), and felt it was wise to leave to God the understanding of His own mystery.

Thus, Ibn Hanbal became a strong proponent of the bi-lā kayfa formula. This mediating principle allowed the traditionalists to deny ta'wil (figurative interpretations) of the apparently anthropomorphic texts while concomitantly affirming the doctrine of the "incorporeal, transcendent deity".

Although he argued for literalist meanings of the Qur'anic and prophetic statements about God, Ibn Hanbal was not a fideist and was willing to engage in hermeneutical exercises. The rise of Imam Ahmad ibn Hanbal and the Ashab al-Hadith, whose cause he championed, during the Mihna; would mark the stage for the empowerment and centering of corporealist ideas.

Ibn Hanbal also recognized "Divine Form (Al-Şūrah)" as a true attribute of God. He disagreed with those speculative theologians who interpreted the Divine Form as something that represents pseudo-divinities such as the sun, moon, stars, etc. For Ibn Hanbal, to deny that God truly has a Form is Kufr (disbelief). He also believed that God created Adam "according to His form". Censuring those who alleged that this was referring to the form of Adam, Ibn Hanbal asserted: "He who says that Allah created Adam according to the form of Adam, he is a Jahmi (disbeliever). Which form did Adam have before He created him?"

With respect to Qur'an, one of Ibn Hanbal's most famous contributions to Sunni thought was the considerable role he played in bolstering the orthodox doctrine of the Quran being the "uncreated Word of God" (kalām Allāh ghayr makhlūķ). By "Quran," Ibn Hanbal understood "not just an abstract idea but the Quran with its letters, words, expressions, and ideas

- the Quran in all its living reality, whose nature in itself," according to Ibn Hanbal, eluded human comprehension.

With respect to taqlid, Ahmad ibn Hanbal favored Ijtihad and rejected Taqlid; the practice of blind adherence to madhabs (fiqhs). Ahmad ibn Hanbal's staunch condemnation of Taqlid is reported in Hanbali Qadhi 'Abd al-Rahman ibn Hassan's treatise (1782-1868 C.E) “Fath al-Majeed”. Comparing Taqlid to Shirk (polytheism), ibn Hanbal states:

> "I am amazed at those people who know that a Sanad (i.e. Chain of Transmission) is authentic and yet, in spite of this, they follow the opinion of Sufyan, for Allah (Glorified be He), says: {And let those who oppose the Messenger's commandment (i.e. his Sunnah - legal ways, orders, acts of worship, statements) (among the sects) beware, lest some Fitnah (disbelief, trials, afflictions, earthquakes, killing, overpowered by a tyrant) should befall them or a painful torment be inflicted on them}. (An-Nur: 63). Do you know what that Fitnah is? That Fitnah is Shirk (polytheism). Maybe the rejection of some of his words would cause one to doubt and deviate in his heart and thereby be destroyed."

With respect to intercession, it is narrated by Abū Bakr al-Marwazī in his Mansak that Ibn Hanbal preferred one to make tawassul or "intercession" through the Prophet in every supplication, with the wording:

> "O God! I am turning to Thee with Thy Prophet, the Prophet of Mercy. O Muhammad! I am turning with you to my Lord for the fulfillment of my need."

This report is repeated in many later Hanbali works, in the context of personal supplication as an issue of fiqh. Ibn Qudamah, for example, recommends it for the obtainment of need in his Wasiyya. In the same way, Ibn Taymiyyah cites the Hanbali fatwa on the desirability of the Prophet's intercession in every personal supplication in his "Qāida fil-Tawassul wal-Wasiīla" where he attributes it to "Imām Ahmad and a group of the pious ancestors" from the Mansak of al-Marwazī as his source.

With respect to Sufiism, as there exist historical sources indicating patently "sufi elements in his personal piety" and documented evidence of his amiable interactions with numerous early Sufi and awliyā', including Maruf Karkhi, it is recognized that Ibn Hanbal's relationship with many of the Sufis was one of mutual respect and admiration.

Qadhi Abu Ya'la reports in his Tabaqat: "[Ibn Hanbal] used to greatly respect the Sūfīs and show them kindness and generosity. He was asked about them and was told that they sat in mosques constantly to which he replied, 'Knowledge made them sit.'" Furthermore, it is in Ibn Hanbal's Musnad that we find most of the hadith reports concerning the abdal, forty major saints "whose number [according to Islamic mystical doctrine] would remain constant, one always being replaced by some other on his death" and whose key role in the traditional Sufi conception of the celestial hierarchy would be detailed by later mystics such as Hujwiri and Ibn Arabi.

It has been reported that Ibn Hanbal explicitly identified Maruf Karkhi as one of the abdal, saying: "He is one of the Substitute- awliyā', and his supplication is answered." Of the same Sufi, Ibn Hanbal later asked rhetorically: "Is religious knowledge anything else than what Maruf has achieved?" Additionally, there are accounts of Ibn Hanbal extolling the early ascetic saint Bishr the Barefoot and his sister as two exceptional devotees of God, and of his sending people with mystical questions to Bishr for guidance. It is also recorded that Ibn Hanbal said, with regard to the early Sufis, "I do not know of any people better than them." Moreover, there are accounts of Ibn Hanbal's son, Sālih, being exhorted by his father to go and study under the Sufis. According to one tradition, Sālih said: "My father would send for me whenever a self-denier or ascetic (zāhid aw mutaqashshif) visited him so I could look at him. He loved for me to become like this."

As for the Sufis' reception of Ibn Hanbal, it is evident that he was "held in high regard" by all the major Sufis of the classical and medieval periods, and later Sufi chroniclers often designated the faqih as a saint in their hagiographies, praising him both for his legal work and for his appreciation of Sufi doctrine. Hujwiri, for example, wrote of him: "He was distinguished by devoutness and piety ... Sufis of all orders regard him as blessed. He associated with great Shaykhs, such as Dhul-Nun of Egypt, Bishr al-Hafi, Sari al-Saqati, Maruf Karkhi, and others. His miracles were manifest and his intelligence sound ... He had a firm belief in the principles of religion, and his creed was approved by all the [theologians]." Both non-Hanbali and Hanbali Sufi hagio-graphers such as Hujwiri and Ibn al-Jawzi, respectively, also alluded to Ibn Hanbal's own gifts as a miracle worker and of the blessedness of his grave. For example, Ibn Hanbal's own body was traditionally held to have been blessed with the miracle of incorruptibility, with Ibn al-Jawzi relating: "When the Prophet's descendant Abū Ja'far ibn Abī Mūsā was buried next to him, Ahmad ibn Hanbal's tomb was exposed. His corpse had not putrified and the shroud was still whole and undecayed."

It is evident that "during the first centuries some major Sufis [such as Ibn Ata Allah, Hallaj, and Abdullah Ansari] ... followed the Hanbalite school of law."

By the twelfth-century, the relationship between Hanbalism and Sufism was so close that one of the most prominent Hanbali faqih, Abdul Qadhir Jilani, was also simultaneously the most famous Sufi of his era, and the Tariqa that he founded, the Qadhiriyya, has continued to remain one of the most widespread Sufi orders up till the present day.

Even later Hanbali authors who were famous for criticizing some of the "deviances" of certain heterodox Sufi orders of their day, such as Ibn Qudamah, Ibn al-Jawzi, and Ibn Qayyim al-Jawziyya, all belonged to Abdul Qadhir Jilani's order themselves, and never condemned Sufism outright.

With respect to relics, as has been noted by scholars, it is evident that Ibn Hanbal "believed in the power of relics," and supported the seeking of blessing through them in religious veneration. Indeed, several accounts of Ibn Hanbal's life relate that he often carried "a purse ... in his sleeve containing ... hairs from the Prophet." Furthermore, Ibn al-Jawzi relates a tradition narrated by Ibn Hanbal's son Abdullah, who recalled his father's devotion towards relics thus: "I saw my father take one of the Prophet's hairs, place it over his mouth, and kiss it. I may have seen him place it over his eyes, and dip it in water and then drink the water for a cure." In the same way, Ibn Hanbal also drunk from the Prophet's bowl (technically a "second-class" relic) in order to seek blessings from it, and considered touching and kissing the sacred minbar of the Prophet for blessings a permissible and pious act. Ibn Hanbal later ordered that he be buried with the hairs of the Prophet he possessed, "one on each eye and a third on his tongue."

The following books of Imam Aḥmad ibn Ḥanbal are found in Ibn al-Nadim's Fihrist:

- Musnad of Imam Ahmad ibn Hanbal
- Usool as-Sunnah : "Foundations of the Prophetic Tradition (in Belief)"

- asSunnah : "The Prophet Tradition (in Belief)"
- Kitab al-`Ilal wa Ma'rifat al-Rijal: "The Book of Narrations Containing Hidden Knowledge of the Men (of Hadeeth)" Riyad: Al-Maktabah al-Islamiyyah
- Kitab al-Manasik: "The Book of the Rites of Hajj"
- Kitab al-Zuhd: "The Book of Abstinence" ed. Muhammad Zaghlul, Beirut: Dar al-Kitab al-'Arabi, 1994
- Kitab al-Iman: "The Book of Iman"
- Kitab al-Masa'il "Issues in Fiqh"
- Kitab al-Ashribah: "The Book of Drinks"
- Kitab al-Fada'il Sahaba: "Virtues of the Companions"
- Kitab Tha'ah al-Rasul : "The Book of Obedience to the Messenger"
- Kitab Mansukh: "The Book of Abrogation"
- Kitab al-Fara'id: "The Book of Obligatory Duties"
- Kitab al-Radd `ala al-Zanadiqa wa'l-Jahmiyya "Refutations of the Heretics and the Jahmites" (Cairo: 1973)
- Tafsir: "Awwalogy"

Biographical Summary

Ahmad ibn Hanbal's family was originally from Basra, Iraq, and belonged to the Arab Banu Dhuhl tribe. His father was an officer in the Abbasid army in Khurasan and later settled with his family in Baghdad, where Ahmad was born in 780 CE.

Ibn Hanbal had two wives and several children, including an older son, who later became a qadhi in Isfahan.

Imam Aḥmad ibn Ḥanbal studied extensively in Baghdad, and later traveled to further his education. He started learning Fiqh under the celebrated Hanafi judge, Abu Yusuf, the renowned student and companion of Imaam Abu Hanifah. After finishing his studies with Abu Yusuf, ibn

Hanbal began traveling through Iraq, Syria, and Arabia to collect hadiths, or traditions of the Prophet Muhammad. Ibn al-Jawzi states that Imam Aḥmad ibn Ḥanbal had 414 Hadith narrators (Ravis) whom he narrated from. With this knowledge, he became a leading authority on the hadith, leaving an immense encyclopedia of hadith, which he wrote under the title "al-Musnad".

After several years of travel, he returned to Baghdad to study Islamic fiqh under another founder of jurisprudence school, Imam Al-Shafi'i.

Aḥmad ibn Ḥanbal became a mufti in his old age, and founded the Hanbali madhab, or school of fiqh, which is now most dominant in Saudi Arabia, Qatar, and the United Arab Emirates. Unlike the other three schools of fiqh (Hanafi, Maliki, and Shafi), the Hanbali madhab remained largely traditionalist or Athari in ilm-al-kalam.

In addition to his scholastic enterprises, ibn Hanbal was a soldier on the Islamic frontiers (Ribat) and made Hajj five times in his life, twice on foot.

Ibn Hanbal passed away on Friday, 12 Rabi-ul-awwal, 241 AH/ 2 August, 855 AD at the age of 74–75 in Baghdad, Iraq.

Historians relate that his funeral was attended by 800,000 men and 60,000 women and that 20,000 Christians and Jews converted to Islam on that day. His grave is located in the premises of the Imam Ahmad Bin Hanbal Shrine in Ar-Rusafa District.

Imprisonment (Mihna): Ibn Hanbal was known to have been called before the Inquisition or Mihna of the Abbasid Caliph Al-Ma'mun. Al-Ma'mun wanted to assert the religious authority of the Caliph by pressuring ulama to adopt the Mu'tazila view that the Qur'an was created rather than uncreated. According to Sunni tradition, ibn Hanbal was among the ulama to resist the Caliph's interference and the Mu'tazili doctrine of a created Qur'an. Ibn Hanbal's stand against the inquisition by the Mu'tazila (who had been the ruling authority at the time) led to the Hanbali school establishing itself firmly as not only a school of fiqh, but of kalam as well.

Due to his refusal to accept Mu'tazilite authority, ibn Hanbal was imprisoned in Baghdad throughout the reign of al-Ma'mun. In an incident during the rule of al-Ma'mun's successor, al-Mu'tasim, ibn Hanbal was flogged to unconsciousness. However, this caused upheaval in Baghdad and al-Mu'tasim was forced to release ibn Hanbal. After al-Mu'tasim's death, al-Wathiq became caliph and continued his predecessor's policies of Mu'tazilite enforcement and in this pursuit, he banished ibn Hanbal from Baghdad. It was only after al-Wathiq's death and the ascent of his brother al-Mutawakkil, who was much friendlier to the more traditional Sunni beliefs, that ibn Hanbal was welcomed back to Baghdad.

56. Dāwūd al-Ẓāhirī

Dāwūd ibn ʿAlī ibn Khalaf al-Ẓāhirī

(Arabic: دَاوُدُ بنُ عَلِيِّ بنِ خَلَفٍ الظَّاهِرِيُّ),

(815–883 CE / 199–269 AH)),

was a Sunnī mutakallim, faqih, and mohaddith specialized in the study of Islamic law (*sharīʿa*) and the fields of awwalogy, ilm-al-rijal, and factology. He was the eponymous founder of the Ẓāhirī fiqh (*madhhab*), the fifth school of thought in Sunnī Islam, characterized by its adherence on the outward (*ẓāhir*) meaning of expressions in the Quran and ḥadīth; the consensus (*ijmāʿ*) of the first generation of Muhammad's closest companions (*ṣaḥāba*) for sources of Islamic fiqh; and rejection of analogical deduction (*qiyās*) and societal custom or knowledge (*urf*). These are used by other fiqhs. He was a celebrated, if not controversial, figure during his time, being referred to in Islamic historiographical texts as "the scholar of the era."

Scientific Contributions

Al-Dhahabī states that al-Ẓāhirī learnt ilm-al-*kalām* from Ibn Kullāb. Similarly to other Muslim scholars who were accused of sharing Ibn Kullāb's creed (*ʿaqīdah*), such as Ḥārith al-Muḥāsibī and Muḥammad al-Bukhārī, al-Ẓāhirī was repudiated by certain factions of *ḥadīth* authorities of his era, which accused him of holding particular creedal views relating to God's speech (Quran).

Ahle Hadith: Al-Ẓāhirī's understanding of the Islamic faith was described by al-Dhahabī's teacher, the Syrian historian and scholar Ibn Taymiyyah, as having been based upon the Atharī *ʿaqīdah*, affirming the attributes of God without delving into their fundamental nature. Muḥammad ibn ʿAbd al-Karīm al-Shahrastānī, a 12th-century Persian

Muslim historian of religion and Ash'arī theologian, classified al-Ẓāhirī along with Mālik ibn Anas (founder of the Mālikī school), Aḥmad ibn Ḥanbal, and Sufyān al-Thawrī as early Sunnī Muslim scholars who rejected both esoteric and anthropotheistic interpretations of God, but both Ibn Taymiyyah and al-Shahrastānī considered al-Ẓāhirī and his students, along with Mālik ibn Anas, al-Shāfi'ī , Ibn Ḥanbal, al-Thawrī, Abū Thawr, al-Māwardī, and their students to be the *Ahl al-Ḥadīth* ("people of the tradition"), as opposed to the *Ahl al-Ra'y* ("people of logic").

Zaharism: This creed of not delving into the fundamental nature of the texts likely affected al-Ẓāhirī's views on literalism. While all the major figures of Islam were united upon the Quran and *sunnah* being the foremost sources of fiqh, al-Ẓāhirī held that these two sources must also be taken at the literal meanings and only applied in the particular circumstances which they described. Use of Qiyas would pollute the decision with human insertion. Therefore, Al-Ẓāhirī rejected the principle of **qiyās**, otherwise known as "analogical reasoning", as a method of deducing rulings in fiqh. Al Zahiri regarding the practice as a form of *bid'ah*, which means "innovation" within the Islamic religion, and this the Islamic prophet Muhammad had not allowed. There are conflicting views regarding al-Ẓāhirī's position when the specific causality of a command or prohibition within the Quran or prophetic example was stated. Some take the view that al-Ẓāhirī restricted the ruling to the incident or condition in which the causality arose, seeing that this causality provides a concrete law. On the other hand, the use of Qiyas would widen the causality given that different ravis and mohadditheen, or the mofassireen in the case of Quran, recorded opposing statements.

For similar reasons, Al-Ẓāhirī considered the scholarly consensus (***ijmā'***) to consist only of the opinions of the first generation of

Muhammad's closest companions (*ṣaḥāba*); excluding all other generations after them from this definition.

Nature of Quran: While al-Ẓāhirī at one time studied the *ḥadīth* literature under Aḥmad ibn Ḥanbal, he was later barred from study due to a dispute regarding the nature of the Quran. Al-Ẓāhirī stated that the Quran was *muhdath* or "recently occurring", a stance of which Ibn Ḥanbal strongly disapproved.

> Even before that time, Ibn Ḥanbal had actually cut off contact with anyone who would study with or consult al-Ẓāhirī regarding religious matters, a habit which Ibn Ḥanbal started after witnessing Ẓāhirī's defense of al-Shāfiʿī against the attacks of Ibn Rāhwayh. The rumor regarding al-Ẓāhirī's statement about the Quran only added more fuel to the fire.

The Syrian historian and scholar Ibn Taymiyyah said that the dispute was semantic in nature, arising from a confusion of al-Ẓāhirī's intended meaning: namely, that God is unique and existent without peers (*tawḥīd*).

Thus al-Ẓāhirī, Ibn Ḥanbal, al-Shāfiʿī, Mālik ibn Anas, al-Thawrī, Ibn Rāhwayh, al-Ṭabarī, Abū Ḥanīfa al-Nuʿmān, ʿAbd al-Raḥmān al-Awzāʿī, Ibn Khuzaymah, ʿAbdullah ibn Mubārak, al-Dārimī, and Muḥammad al-Bukhārī - described by Ibn Taymiyyah as the leading figures of Islam at the time - all agreed that the Quran was uncreated, but a semantic misunderstanding arose when al-Ẓāhirī, al-Bukhārī, Muslim bin al-Ḥajjāj, and others used the phrase "Quran was *muhdath* " to establish that God and the Quran (believed by Muslims to be the literal speech of God) are not the same thing, but rather that God's speech is an attribute.

Modern-day scholarship has suggested, in light of the weakness in the chains of narration connecting the phrase "the Quran is recently occurring" to al-Ẓāhirī that he may have never made such a statement or held such a belief at all. Due to al-Ẓāhirī's denial of analogical reasoning and blind following - cornerstones in the other main Sunnī schools of thought - the students of those schools may have forged the statement and attributed it to al-Ẓāhirī, as a means of pushing the common people away from him and his eponymous school of thought. Abū ʿUbaida further supported his point by noting that al-Ẓāhirī and his students were actually severer in their opposition to the Muʿtazilite school and their belief that the Quran was created, than Ibn Ḥanbal was; the Zaharis used harsh language in their written responses to Mutazillites.

Usury: Al-Ẓāhirī held the view that regarding in-kind exchanges of goods, the forbidden type of usury applies only to the six commodities specified by the Islamic prophet Muhammad: gold, silver, wheat, barley, dates, and salt. Because al-Ẓāhirī rejected the use of analogical reasoning in fiqh, he disagreed with the majority view that the prohibition on excess gain in in-kind exchanges is for all commodities; and al-Zahiri did not consider such gains to be a form of usury. Had Muhammad intended to include commodities other than the above six, he could have done so; because he specified that usury was only prohibited in these six commodities and that Muslims were free to deal in other commodities as they liked, al-Ẓāhirī saw no basis for making an analogy to any other commodities.

Face Veil: According to Muḥammad ash-Shawkānī, al-Ẓāhirī regarded the Muslim face veil to be recommended (*mustaḥabb*) rather than obligatory (*wajīb*), seeing that a woman's face could be uncovered in public but that all other body parts must be covered. This was the position of Ibn Ḥanbal as well.

Travel: If a Muslim begins traveling while fasting (*ṣawm*) during the month of Ramaḍān, al-Ẓāhirī saw that the individual should break their fast on the day which they started their journey, a view upon which both Ibn Ḥanbal and Ibn Rāhwayh agreed. This was due to the Quranic verse allowing the traveler to skip the Ramaḍān fast and make it up when they complete their journey. If a Muslim did fast while traveling, they would still have to make up for it, according to al-Ẓāhirī's view, as the verse wasn't merely an allowance for breaking the fast, but a command.

Most Muslims shorten the length of their prayers while traveling. This "traveling" by which the Muslim shortens his prayers and breaks the fast is a topic of discussion among jurists, and they consider the distance and duration of travel. Al-Ẓāhirī saw that any form of traveling, regardless of distance or duration, allowed the individual to shorten their prayers.

Al-Zahiri's Writings: Al-Ẓāhirī was known as being a prolific author, and the Arab-Persian historian and bibliographer. Ibn al-Nadīm was able to personally record the names of at least 157 of his written works, the majority on topics within Islamic studies. Some of these works were very long, and they covered both legal theory and all branches of law. He was also considered to be the first person to have written a biography of his former teacher, al-Shāfiʿī. None of these works have survived to the modern era in their entirety.

Ibn al-Nadīm also mentions that after al-Shāfiʿī's treatise *Al-Risala*, Ibn Ḥanbal and al-Ẓāhirī were the next major ulama to author works on the principles of Islamic jurisprudence (*Uṣūl al-Fiqh*). Al-Ẓāhirī produced a number of works on various topics, including his rejection of blindly following the Islamic clergymen; the difference between general and specific verses of the Quran; the difference between succinct and detailed commands in the Islamic religion; and his views on and experiences with his former teacher, al-Shāfiʿī. Modern ulama have pieced together chapter headings for al-Ẓāhirī's work on juristic principles from other early works. These chapters are in the following order:

- binding consensus,
- invalidity of blindly following the clergy,
- invalidity of analogical reasoning,
- traditions transmitted by single authorities,
- traditions which provide certainty,
- incontrovertible proof,
- particular vs. general scriptural texts, and
- specified vs. unspecified texts.

The chapters, and perhaps even the information contained therein, have primarily been preserved in the Fāṭimid-era works of the Ismāʿīlī Shīʿīte faqih, Qāḍī al-Nuʿmān, in addition to the passages preserved in the treatise *Al-Muhalla* of the Sunnī Muslim historian Ibn Ḥazm, an adherent of the Ẓāhirīte school.

Although some of al-Ẓāhirī's views in kalam and fiqh were controversial, his character and religious piety carry universal acclaim. The Muslim scholars al-Khaṭīb al-Baghdādī, al-Dhahabī, al-Ṭabarī, al-Nawāwī, al-Suyūṭī, and al-Albānī all attested to his morality, humility, and personal ethics.

Al Zahiri and other Ulama: While the Ẓāhirīte school is not as numerous today as the other four major Sunnī schools of fiqh, it was once a major school and encompassed Mesopotamia, the Iberian Peninsula, the Balearic Islands, North Africa, and Southern Iran. Even his contemporary critics conceded to his intellect and level of knowledge, though they rejected some of his beliefs. He has been described as "the scholar of the era" by al-Dhahabī, and the hierarchy of religious knowledge in Baghdad was considered to have ended with al-Ẓāhirī at the top. When al-Ṭabarī was asked regarding the books of Ibn Qutaybah, he answered that Ibn Qutaybah's work was "nothing", and recommended the books of the "people of fiqh", mentioning al-Shāfiʿī and al-Ẓāhirī by name, then "their contemporaries".

Members of other schools have often criticized al-Ẓāhirī for his rejection of analogical reasoning. The early followers of al-Shāfiʿī in general held negative views of their former classmate, and the followers of the Shāfiʿīte school, al-Juwāynī in particular, were harsh upon al-Ẓāhirī himself. However, many followers of the Shāfiʿī school have taken more accommodating views of al-Ẓāhirī's legal rulings. Al-Dhahabī defended al-Ẓāhirī and his followers, stating that just as al-Juwāynī had arrived to his views by the process of scholarly discourse, so had al-Ẓāhirī. Likewise, Ibn al-Ṣalāḥ also defended the legitimacy of al-Ẓāhirī's views and his school, listing a number of figures from other Sunnī schools of thought who considered al-Ẓāhirī's opinions as scholarly discourses.

Shīʿa Muslims have taken a dimmer view of al-Ẓāhirī and his school.

The Ismāʿīlī Shīʿīte jurist Qāḍī al-Nuʿmān was critical of al-Ẓāhirī for rejecting qiyas.

Being steeped in esoteric philosophy, *the Muʿtazila* school were quite hostile towards al-Ẓāhirī and his school. Although some prominent figures of this school, such as the Muʿtazilite theologian Ibrāhīm al-Naẓẓām, denied the validity of analogical reasoning as al-Ẓāhirī did, they also denied literalism and the validity of consensus (Ijma), and most of them found al-Ẓāhirī's ideas to be ridiculous.

Biographical Summary

Al-Ẓāhirī's father was Arab whereas his mother was likely Persian. Historians and scholars Ibn Ḥazm, al-Dhahabī, Christopher Melchert and Ignác Goldziher, hold that he was of Iraqi origins, having been born in the city of Kufa. His father was in the civil service of the Abbasid caliph al-Maʿmūn, stationed in Kashan, a smaller city near Isfahan.

During his formative years, al-Ẓāhirī relocated from Kufa to Baghdad and studied the tafsir of *ḥadīth* and Quranic with a number of notable scholars of the time, including Abū Thawr, Yaḥyā ibn Maʿīn, and Aḥmad ibn Ḥanbal. His study under renowned figures of *Atharī* kalam was in contrast to the views of his father, who was a follower of the less orthodox Ḥanafī school.

Indian alim Chiragh Ali has suggested that Ẓāhirī's fiqh was, like that of Ibn Ḥanbal, actually a direct reaction to the Ḥanafī fiqh.

Toward the end of his education, al-Ẓāhirī traveled to Nishapur in Greater Khorasan in order to complete his studies with Isḥāq ibn Rāhwayh, at the time considered a champion of the Sunnī philosophy. Ibn al-Jawzī states that, when studying with Ibn Rāhwayh, considered one of the most knowledgeable ulama, al-Ẓāhirī was willing to debate with Ibn Rāhwayh on religious topics, something no one else had ever dared to do.

Ibn Rāhwayh criticized Muḥammad ibn Idrīs al-Shāfiʿī, founder of the Shāfiʿī school, during one of his lessons; a debate ensued in which al-Ẓāhirī alleged that Ibn Rāhwayh didn't understand al-Shāfiʿī 's point on the topic of discussion. Aḥmad ibn Ḥanbal was physically present for the debate, and he declared Ibn Rāhwayh to be the winner.

Al-Ẓāhirī was initially a follower of al-Shāfiʿī in matters of fiqh, later branching off in terms of his principles, likely due to the influence of Ibn Rāhwayh.

After completing his studies in Nishapur, al-Ẓāhirī returned to Baghdad and began delivering his own lessons. While historians differ regarding his exact number of students, it is agreed that his following was large, with most estimates ranging between four and five hundred students who would regularly attend his *majlis.* His reputation spread outside of Baghdad, and even high-level scholars from elsewhere in the world began seeking al-Ẓāhirī's advice on topics of religious study. While his views were not universally accepted in his time, no attempts were made by his contemporaries to prevent him from granting religious verdicts, nor were they opposed to his teaching position. His most well-known students were his son Abū Bakr Muḥammad ibn Dāwūd al-Iṣfahānī; ʿAbdullah, the son of Aḥmad ibn Ḥanbal; and al-Ṭabarī, Nifṭawayh, and Ruwaym. Al-Ẓāhirī was also the teacher of the Sunnī Muslim jurist ʿAbd Allāh al-Qaysī, who was responsible for spreading the Ẓāhirīte school in Al-Andalus.

Al-Ẓāhirī died during the month of Ramaḍān in Baghdad, where he was buried. Historians have stated both 883 CE and 884 CE as the year of his death.

57. Ibn Hazm

Abū Muḥammad ʿAlī ibn Aḥmad ibn Saʿīd ibn Ḥazm

(Arabic: أبو محمد علي بن احمد بن سعيد بن حزم)

(also sometimes known as al-Andalusī aẓ-Ẓāhirī),

(7 November 994 – 15 August 1064),

was a polymath, historian, faqih, qadhi, philosopher, and mutakallim; born in the Caliphate of Córdoba, Andalusia.

Scientific Contributions

Described as one of the strictest hadith interpreters, Ibn Hazm was a leading proponent and codifier of the Zahiri school of Islamic thought and authored 400 works, of which only 40 still survive. In all, his written works amounted to some 80 000 pages. Much of Ibn Hazm's substantial body of works, which approached that of Muhammad ibn Jarir al-Tabari and As-Suyuti's, was burned in Seville by his sectarian and political opponents. His surviving works, while criticized as repetitive, didactic and abrasive in style, also show a fearless irreverence towards his academic critics and authorities.

Ibn Hazm wrote works on fiqh and kalam and over ten medical books. He called for science to be integrated into a standard curriculum. In Organization of the Sciences, he diachronically defines educational fields as stages of progressive acquisition set over a five-year curriculum, from language and awwalogy of the Qur'an to the life and physical sciences to a rationalistic kalam.

In Fisal (Detailed Critical Examination), a treatise on Islamic science and kalam, Ibn Hazm promoted sense perception above subjectively flawed human reason. Recognizing the importance of reason, as the Qur'an itself invites reflection, he argued that reflection to refer mainly to revelation and sense data since the principles of reason are themselves derived entirely from sense experience. He concludes that reason is not a faculty for independent research or discovery, but that sense perception should be used in its place, an idea that forms the basis of supremacy of experimental observations over mere rational thinking without supporting measurements. This is the idea centuries ahead of its times.

Perhaps ibn Hazm's most influential work, selections of which have been translated into English, is now The Muhalla (المحلى بالأثار), or The Adorned Treatise. It is reported to be a summary of a much longer work, known as Al-Mujalla (المجلى). Its essential focus is on matters of fiqh (فقه), but it also touches on matters of aqida in its first chapter, Kitab al-Tawheed (كتاب التوحيد), whose focus is on matters related to monotheism and the fundamental principles of approach to divine texts. One of the main points that emerges from the masterpiece of thought in fiqh is that Ibn Hazm rejects analogical reasoning (qiyas قياس) but prefers a far more direct and literal approach to the texts.

> Out of the six sources of fiqh processes (Quran, Hadith, Ijma, Qiyas, Istihsan, Urf) that Abu Hanifa established, Ibn Hazm recognizes only two: Quran and Hadith. *The root causes of the conflicts between the Madhabs* are generally due to the expansive landscape that the four Imams introduced for their fiqh *convenience*, together with the practice of *Taqlid* that the latter day Ulema established.

Ibn Hazm wrote the Scope of Logic, which stressed on the importance of sense perception as a source of knowledge. He wrote that the "*first sources of all human knowledge are the soundly used senses and the intuitions of reason, combined with a correct understanding of a language*". Ibn Hazm also criticized some of the more traditionalist mutakallimeen who were opposed to the use of logic and argued that the first generations of Muslims did not rely on logic. His response was that the early Muslims had witnessed the revelation directly, but later Muslims have been exposed to contrasting beliefs and so *the use of logic is necessary to preserve the true teachings of Islam.* The work was first republished in Arabic by Ihsan Abbas in 1959 and most recently by Abu Abd al-Rahman Ibn Aqil al-Zahiri in 2007.

In his book, In Pursuit of Virtue, Ibn Hazm had urged his readers:

> Do not use your energy except for a cause more noble than yourself. Such a cause cannot be found except in Almighty God Himself: to preach the truth, to defend womanhood, to repel humiliation which your creator has not imposed upon you, to help the oppressed. Anyone who uses his energy for the sake of the vanities of the world is like someone who exchanges gemstones for gravel.

Ibn Hazm's teachers in medicine included al-Zahrawi and Ibn al-Kattani, and he wrote ten medical works, including Kitab fi'l-Adwiya al-mufrada mentioned by al-Dhahabi.

Biographical Summary

Ibn Hazm's grandfather Sa'id and his father, Ahmad, both held high advisory positions in the court of Umayyad Caliph Hisham II. Scholars believe that they were Iberian Christians who converted to Islam.

Having been raised in a politically and economically important family, Ibn Hazm mingled with people of power and influence all his life. He had access to levels of government by his adolescence that most people then would never know throughout their whole lives. Those experiences with government and politicians caused Ibn Hazm to develop a reluctant and even sad skepticism about human nature and the capacity of human beings to deceive and to oppress.

His reaction was to believe that there was no refuge or truth except with an infallible God and that with men resided only corruption. He was thus known for his cynicism regarding humanity and a strong respect for the principles of language and sincerity in communication.

Ibn Hazm lived among the circle of the ruling hierarchy of the Caliphate of Córdoba government. His experiences produced an eager and observant attitude, and he gained an excellent education at Córdoba.

After the death of the grand vizier, al-Muzaffar, in 1008, the Caliphate of Iberia became embroiled in a civil war that lasted until 1031 and resulted in its collapse of the central authority of Córdoba and the emergence of many smaller incompetent states, the taifas.

Ibn Hazm's father died in 1012. Ibn Hazm was frequently imprisoned as a suspected supporter of the Umayyads. By 1031, Ibn Hazm retreated to his family estate at Manta Lisham and had begun to express his activist convictions in the literary form. He was a leading proponent and codifier of the Zahiri school of Islamic thought, and he produced a reported 400 works, but only 40 still survive. His political and religious opponents gained power after the collapse of the caliphate and so he accepted an offer of asylum from the governor of the island of Majorca in the 1040s. He continued to propagate the Zahiri School there before he returned to Andalusia.

Contemporaries coined the saying "the tongue of Ibn Hazm was a twin brother to the sword of al-Hajjaj", an infamous 7th century general and governor of Iraq. Ibn Hazm became so frequently quoted that the phrase "Ibn Hazm said" became proverbial.

As an Athari, he opposed the allegorical interpretation of religious texts and preferred a grammatical and syntactical interpretation of the Qur'an. He granted cognitive legitimacy only to revelation and sensation, and he considered deductive reasoning insufficient in legal and religious matters. He rejected practices common among more orthodox schools, such as qiyas and istihsan. He was initially a follower of the Maliki school of law within Sunni Islam, but he switched to the Shafi'i. Around the age of thirty, he finally settled with the Zahiri School. He is perhaps the most well-known adherent of the school and the main source of extant works on Zahirite fiqh. He studied the school's precepts and methods under Abu al-Khiyar al-Dawudi al-Zahiri of Santarém Municipality and was eventually promoted to the level of a teacher of the school himself. In 1029, both were expelled from the main mosque of Cordoba for their activities.

Ibn Hazm was born on 7 November 994, and he died on 15 August 1064.

Ruhaniyaat

Spirit and soul are not Islamic; they are Christian concepts. Their use by the translators, are sometimes misleading due to their Christian content and baggage.

The one word that Quran uses is Ruh, which is sometimes translated as the "breath". That again is simple minded on the subject of ruh, for Quran itself declares that it is a secret among the secrets.

Any experience of the ruh is referred to as Ruhani experience. The act of seeking such an experience is called Ruhaniyat. Its plural is Ruhaniyaat, which also means the science of ruh.

The term represents the science of Islamic seeking of the truth about life, based on the practices of Salat (prayer), Saum (fasting) and Zikr (recall). These people are Scientists of Ruhaniyyaat. Non-Muslim investigators like the Christian Saints might also be candidates if they meet the criteria. However, Muslims enjoy far greater freedom from the central religious authority, which does not exist in Islam or among Muslims. Unlike Christian pursuit of spirituality, Muslims practice ruhaniyyaat freely, without coercion from a central authority.

Muslim "seekers" have therefore been able to make ruhaniyyaat into a genuine science with detailed methodologies and verification markers along the path, technically called the Tariqa.

On the other hand, Christians generally think that it is from Grace and not from Deeds, and thus the path (Tariqah) becomes almost irrelevant for Christians. Further, the Church confers Sainthood by its own initiative and criteria. And it happens after the death of the candidate. This is so unlike a science, so that Christian Saints do not generally qualify as seekers of Ruhaniyyaat, or Sufis.

The goal of ruhaniyyaat is to know the **Haqiqah** which signifies the reality of life. There is a nominal way to proceed in search of Haqiqah, and this way is called *Tariqa*. One essential part of Tariqa is to purify the inner self of one's own self. This process is known as *Tazkiyah*. One way for Tazkiyah is to practice *Zuhd*, which means, in part, that the seeker avoids luxury in life and adopts a type of detachment from the worldly aspects of life in order to establish a strong attachment with Allah: while of course adhering to Salat, Saum and Zakat as integral to Ruhaniyyaat.

Quran declares that ruh is a decision among the "Decisions" of God, and what knowledge of it can be given to people is but a little bit.

Every one concentrates on Salat, Saum and Zakat. For most people this practice is an end in itself, because they feel this practice can afford them salvation in the eyes of God. Sufis also concentrate on the same practices. The Sufis, however, tend to use the practices of Saum, Salat, and Zakat as a means to know the "Decisions of God", even if only a little bit. Doing so sets the Sufis on a different but parallel and equivalent path to most Muslims who make Salat, Saum and Zakat as goals within themselves. It is sometimes stated that the Sufi approach allows a deeper experience with the fundamentals, namely the practice of Salat, Saum and Zakat.

Sufiism is a science the counterpart of which does not exist in English language, nor in Christianity. Even though Christianity has Saints, they are not equivalent to Sufis for two main reasons. First, no science like Sufiism exists within Christianity, for example, no Tariqah is rationally devised for Sainthood: on the contrary, some Christians profess that it is by grace, not by deeds. Second, the Saintship is awarded, posthumously, by the Church on its own initiative and discretion.

58. Hasan al-Basri

Abu Sa'id ibn Abi al-Hasan Yasar al-Basri

(for short Hasan al-Basri),

(Arabic: الحسن البصري),

(642 CE - 15 October 728),

was a faqih, alim, zahid, mutakallim, and awwalagist.

Scientific Contributions

The particular disciplines in which he is said to have excelled included tafsīr of the Quran, whence his name is invariably encountered in commentaries on the scripture, as well as in ilm-al-kalam.

Very few of Hasan's original writings survived. His proverbs and maxims on various subjects have been transmitted primarily through tradition by his numerous students. Fragments of his sermons have survived in the works of later authors. works that bear his name are apocryphal. Examples of such apocryphal works are the Risālat al-qadar ilā ʿAbd al-Malik (Epistle to ʿAbd al-Malik against the Predestinarians), a pseudepigraphical text from the ninth or early-tenth century; and another is a letter "of an ascetic type" which is of hortatory character, addressed to Umar II (d. 720), which is likewise deemed spurious.

Traditionally, Hasan has been commemorated as an outstanding figure by all the Sunni schools of thought, as he preceded all the Sunni fuqaha. He was frequently designated as one of the well respected of the early Islamic community in later writings by such important Sunni thinkers as Abu Talib al-Makki (d. 996), Abu Nu`aym (d. 1038), Ali Hujwiri (d. 1077), Ibn al-Jawzi (d. 1201), and Attar of Nishapur (d. 1221). In his famed Ḳūt al-ḳulūb, the most important work of Basran ruhaniyyaat, Abu Talib al-Makki says of Hasan: "Ḥasan is our Imām in this doctrine which we represent. We walk in his footsteps and we follow his ways and from his lamp we have our light". There are traditions which relate that some of Hasan's contemporaries did indeed identify him as one of the abdal of that period, though the traditions may be apocryphal, initiated by latter day zealot Sufis.

As a young man, Hasan took part in the campaigns of conquest in eastern Iran (ca. 663) and worked as a jewel-merchant, prior to forsaking the business and military life for that of a faqih, mutakallim, and mohaddith.

It was during this latter period that he began to criticize the policies of the governors in Iraq, even stirring up the authorities to such a degree that he actually had to flee for the safety of his life under the reign of Ḥajjāj, whose anger Hasan had roused due to his forthright condemnation of Ḥajjāj's founding of Wāsiṭ in 705. Hasan began to publicly denounce the accumulation of riches by the wealthy; and it is said that he personally despised wealth. It is reported that he "rejected a suitor for his daughter's hand who was famous for his wealth, simply because of his riches."

Hassan al-Basri is regarded by the Sufi circles as being amongst early Sufis. Many Sufis include him in their sufi lineage. There are many sufi stories about the encounters between Hassan al-Basri and Rabia al-Basri. However, looking at their dates it appears unlikely that those encounters took place as Rabia was of the order of ten years in age when Hassan al-Basri died. However, those stories are good teaching stories in sufi teaching, not to be confused with historicity.

Biographical Summary

Hasan was born in Medina in 642 CE. His mother, Khayra, is said to have been a maidservant of one of Muhammad's wives, Umm Salama (d. 683), while his father, Peroz, was a Persian slave who originally hailed from southern Iraq. According to tradition, Hasan grew up in Medina for the vast portion of his early life, prior to his family's move to Basra after the Battle of Siffin. According to some scholars, it is "primarily this association with Medina and his acquaintance there with many of the notable Companions and wives of Muḥammad that elevated his importance as an authoritative figure in religious and historical context. He became one of the most celebrated tābiʿin, enjoying remarkable posthumous legacy as an alim, mofassir, mohaddith, and zahid.

The various extant biographies relate that Hasan was once nursed by Umm Salama, and that his mother took him after his birth to the caliph Umar (d. 644), who is related to have blessed him with the prayer: "O God! Please do make him wise in the faith and beloved to all people." As an adolescent, he also studied with Imam Ali.

Hasan died in Basra in 728, being eighty-six years old. According to a tradition quoted by the medieval traditionist Qushayri (d. 1074), "on the night of al-Hasan al-Basri's death ... [a local man] saw in a dream that the Gates of Heaven were opened and a crier announced: 'Verily, al-Hasan al-Basri is coming to God Most High, Who is pleased with him.'" However, it is difficult to verify such traditions, and sometimes they are stated as mere stories to make a teaching point.

According to various historical sources, it is said that Hasan was admired by his contemporaries for his handsome appearance. With some asserting he had blue eyes. In this connection, Ibn Qayyim al-Jawziyya (d. 1350) relates an older tradition, which states: "A group of women went out on the day of Eid and went about looking at people. They were asked: 'Who is the most handsome person you have seen today?' They replied: 'It is a teacher wearing a black turban.' They meant al-Ḥasan al-Baṣrī." As for his personality, it is related that Hasan was a frequent weeper, being known by those around him "for the abundance of tears he shed out of compunction for his sins." One particular tradition relates that he wept so much praying on his rooftop one day that his abundant tears began to run off "through the downspouts upon a passerby, who inquired whether the water was clean." Hasan immediately called out to the man below, telling him "it was not, for these were sinner's tears." As such, "he advised the passerby to wash himself forthwith." In a similar vein, Qushayri related of Hasan: "One would never see al-Hasan al-Basri without thinking that he had just been afflicted with a terrible tragedy." With regard to these traditions, one scholar noted that it is evident that Hasan "was deeply steeped in the sadness and fear so typical of ascetics of all religions."

59. Muhammad Ibn Wasi' Al-Azdi

Muhammad Ibn Wasi' Al-Azdi

(d.ca.744 or 751),

was a tabi'i mohaddith, qadhi, and mojahid who was noted for his zuhd. His statement, 'I never saw anything without seeing Allah therein' was much discussed by later Sufis.

Scientific Contrbutions

Muhammad Ibn Wasi' Al-Azdi did jihad under Qutaybah Ibn Muslim (d.715) during the Ummayad conquest of Transoxiana, and later became a qadhi.

There is a (teaching) story that claims that a Muslim saw in a dream that Malik Bin Deenar and Ibn Wasi were being led into Jannah, and Malik was more honored and allowed to enter first. When Malik enquired, noting that he believed Ibn Wasi' was the more noble, he was told that:

> it was true, "but Mohammed ibn Wasi possessed two shirts, and Malik only one. That is the reason why Malik is preferred".

Qutaybah Ibn Muslim said of him, "That the finger of Muhammad ibn Wasi' points to the sky in battle is more beloved to me than one hundred thousand renowned swords and strong youths."

Abu Hamid Al-Ghazali (d.1111) also mentioned him in his writings: If a man finds himself sluggish and averse from austerity and self-discipline, he should consort with one who is a proficient in such practices so as to catch the contagion of his enthusiasm.

One saint used to say, "When I grow lukewarm in self-discipline, I look at Muhammad Ibn Wasi, and the sight of him rekindles my fervor for at least a week."

Muhammad ibn Wasi said; "Only three things do I wish for in this world; a brother to set me straight if I go crooked; a livelihood for which I do not have to beg; and a congregational prayer in which I am relieved of absent mindedness and, which is recorded in my favor.

Biographical Summary

Muhammad ibn Wasi Al-Azdi died in 744 or 751 AD.

60. Rabi'a al-Basri

Rābiʿa al-Baṣrī

(Arabic: رابعة البصري),

(c. 718 - 801),

was a Sufi.

Scientific Contributions

Often noted as having been the single most famous and influential women of zuhd in history, Rābiʿa was renowned for her excelling virtue and tazkiyah. Devoted to zuhd, when asked why she performed a thousand ritual prostrations both during the day and at night, she answered:

"I desire no reward for it; I do it so that the Messenger of God, may God bless him and give him peace, will delight in it on the day of Resurrection and say to the prophets, 'Take note of what a woman of my community has accomplished'".

She was intense in her devotion and love to God. She never claimed to have achieved unity with Him; instead, she dedicated her life to getting closer to God. She used to say:

> "Were the world the possession of a single man, it would not make him rich ... [B]ecause it is passing away."

She was the one who first set forth the doctrine of Divine Love known as Ishq-e-Haqeeqi. She is widely considered being the most important of the early people dedicated to zuhd; a mode of tazkiyah that would eventually become part of Sufi Tariqahs.

It is mythically stated that after a life of zuhd, she spontaneously achieved a state of haqiqah. When asked by Shaikh Hasan al-Basri, how she discovered the secret, she responded by stating: "You know of the how, but I know of the how-less."

The stories detailing the life and practices of Rabi'a al-Adawiyya show her role as superior in ruhani and hikmah matters. In a Sufi narrative, Sufi leader Hasan al-Basri explained, "I passed one whole night and day with Rabi'a ... it never passed through my mind that I was a man nor did it occur to her that she was a woman...when I looked at her I saw myself as bankrupt and Rabi'a as truly haqiqah."

She decided to stay celibate in order to leave her womanhood behind and devote herself completely to God.

Mythical anecdote describes that she was seen running through the streets of Basra carrying a pot of fire in one hand and a bucket of water in the other. When asked what she was doing, she said, "I want to put out the fires of hell, and burn down the rewards of paradise. They block the way to Allah. I do not want to worship from fear of punishment or for the promise of reward, but simply for the love of Allah."

We observe that these mythical stories about Rab'a Basri are teaching stories in Sufiism, perhaps narrated after she had died; these should not be taken as historical.

Following are some present day references to the life of Rabi'a Basri.

- Kayaalp, Pinar, "Rabi'a al-'Adawiyya", in Muhammad in History, Thought, and Culture: An Encyclopedia of the Prophet of God (2 vols.), edited by C. Fitzpatrick and A. Walker, Santa Barbara, ABC-CLIO, 2014, Vol. II, pp. 511–12; ISBN 1610691776
- Mohammad, Shababulqadri Tazkirah e Hazrat Rabia Basri, Mushtaq Book Corner, 2008.

- Rkia Elaroui Cornell, Rabi'a From Narrative to Myth The Many Faces of Islam's Most Famous Woman Saint, Rabi'a al-Adawiyya (Oneworld: London, 2019).
- The life of Rabia has been the subject of several motion pictures by Turkish cinema. One of these films, Rabia, released in 1973, was directed by Osman F. Seden, and Fatma Girik played the leading role of Rabia.
- Rabia, İlk Kadın Evliya (Rabia, The First Woman Saint), another Turkish film on Rabia, also released in 1973 was directed by Süreyya Duru and starred by Hülya Koçyiğit.
- Rabia's quote became song in Indonesia, called "Jika Surga dan Neraka tak pernah ada" sung by Ahmad Dhani and Chrisye in Senyawa Album 2004.
- The final episode of the comedy show, The Good Place, refers to 8th century Sufi mystic poet Hazrat Bibi Rabia Basri, as one of the many worthies who gets into heaven.

Biographical Summary

Rābiʻa is said to have been born between 714 and 718 CE (95 and 98 Hijri) in Basra, Iraq, of the Qays tribe. Farid ud-Din Attar, a later Sufi and poet, recounted much of her early life.

She was the fourth daughter of her family and so named Rābiʻa, meaning "fourth".

According to Fariduddin Attar, when Rābiʻa was born, her parents were so poor that there was no oil in the house to light a lamp, nor even a cloth to wrap her with. Her mother asked her husband to borrow some oil from a neighbor, but he had resolved in his life never to ask for anything from anyone except God. He pretended to go to the neighbor's door and returned home empty-handed. At night Muhammad appeared to him in a dream and told him,

"Your newly born daughter is a favorite of Allah, and shall lead many Muslims to the right path. You should approach the Amir of Basra and present him with a letter in which should be written this message: 'You offer Durood to the Holy Prophet one hundred times every night and four hundred times every Thursday night. However, since you failed to observe the rule last Thursday, as a penalty you must pay the bearer four hundred dinars'".

After the death of her father, famine overtook Basra. She parted from her sisters. Rabia went into the desert to pray and practice zuhd. She was known for her complete devotion as "pure unconditional love of Allah." As an exemplar among others devoted to God, she provided a model of mutual love between Allah and His creation.

Rābiʻa died in her 80s in Basra in 185 AH/801 CE, where her tomb is outside the city.

61. Sari al-Saqati

Sari ibn al-Mughallis al-Saqati

(Arabic: السري بن المغلس السقطي),

(772 – after 833),

was one of the early Muslim Sufi saints of Baghdad. He was one of the most influential students of Maruf Karkhi and one of the first to present Sufism (tasawwuf) in a systematic way.

He was also a friend of Bishr al-Hafi. He was the maternal uncle and spiritual master of Junayd of Baghdad.

Scientific Contributions

Al-Saqati was the shaykh of prominent sufis of his time such as Junayd al-Bahdadi, Abu Said al-Harraz, Abu al-Husayn al-Nuri, Samnun bin Hamza and Ibn Masruq of Baghdad and Khorasan, and Ali al-Gada'iri and Ismail bin Abdullah al-Shami of Syria.

Sulami says that most of the later Sufis followed the way of Sari. Abu Nuaym al-Isfahani and Fariduddin Attar wrote about him as a Sufi with knowledge, wisdom, love, ingenuity and compassion.

According to Sari al-Saqati, being sufi means the following:

- Known for his zuhd and taqwa, Sari al-Sakati was sensitive about and avoided eating and using things whose halal status was questionable.
- He strongly condemned those who made religion a means of livelihood.
- He would advise to choose seclusion (khalwa).
- He would advise those who were engaged in commerce not to separate their hearts from the Truth (al-Haqq) even for a moment, that they should make a living with manual labor and that the divine

light would not reflect on the heart of anyone whose food was doubtful.

- He was so observant of manners in front of Allah that he would not lie on his back or put his feet in the direction of the qibla.
- He was quite humble, always watching himself by avoiding falling into sin, and avoided hypocrisy to the extent that he wished to die in a place where no one knew him, fearing that the earth would not accept his body after his death. The sources state that he was a Sufi who preferred others to himself, saying, "I wish everyone would feel relieved even if I suffered from them". As a matter of fact, according to a legend, when he heard that his own shop was not burned after a fire in the bazaar, he was grateful to Allah saying, "Alhamdulillah", but because he did not share the sorrows of those whose shops were burned, he realized that he had made a mistake and begged God's forgiveness.
- Sari al-Sakati was of the opinion that it is necessary for disciples to learn hadith before they join the path of zuhd and ruhaniyyat, otherwise religious life would be dragged into slackness. He prayed for his disciple Junayd saying, "May Allah grant you to be a person who learns not first tasawwuf and then hadith, but first hadith and then tasawwuf."
- However, like Bishr al-Hafi, he gave importance to understanding the meaning of the hadith rather than narrating hadith, so he did not narrate many hadiths. According to Sari, being a person of zeal in the way of sunnah is better than many deeds done in bid'ah.
- In his understanding of mysticism, observance of the shar'ia and outwardly rules is essential. Knowledge is only valuable to the extent that it leads to action, and the light of knowledge attained by the

sufi should not extinguish the light of taqwa in him. The sign of having knowledge about Allah is to observe His laws and to prefer them to the nafs as much as possible.

- Sari describes the 'arif being like the sun which illuminates every place, the earth that bears the burden of everyone, the water that is the source of life, and the torch that illuminates all sides.
- According to him, esoteric knowledge that contradicts the apparent meanings of the Qur'an and hadiths is invalid. In order to achieve esoteric knowledge, the soul must be taken into account and trained by doing a lot of work.
- Sari al-Saqati said that karamat should not be trusted and given much importance, that the sufi's possession of them would cause them to tear the veils of Divine privacy. He stated that if they cause peace of mind, it means one is being captive to karamat, and he described "istidraj" as "blindness in seeing the faults of the soul".
- Love of God has an important place in his mystical experience, accordingly Sari said, "If the people of love were hit with a sword in their face, they would not know about it"; "My God! Whatever you punish me with just don't torment us by putting a curtain between us."
- According to Sari, tasawwuf is attaining good morals, fulfilling the fard, avoiding the haram, not being heedless, giving khairat and sadaqat, being repentant, being compassionate, as well as not to hurt people, to endure the oppression of the people without holding grudges and thinking about revenge.

Sari al-Sakati was influential on the Sufis after him via educating Junayd of Baghdad, known as "Sayyid al-Taifa", the founder of the Baghdad Sufi school.

Biographical Summary

He was born in 155 AH (772 CE) in the Karkh district of Baghdad. He made a living by continuing his father's scrap business (saqati). In the first period of his life, he traveled as far as Mecca to collect hadith. His master Maruf Karkhi, and Habib al-Ajami (al-Rai), had an influence on his taking the path of tasawwuf.

According to anecdotes, Maruf Karkhi came to Sari's shop with an orphan child asking him to dress the child. Sari fulfilled the request, and with the blessings of the prayer he received from Karkhi, he entered the path of zuhd. He gave 10 coins to Habib al-Ajami to be spent on dervishes. Upon al-Ajami's prayer, his heart turned cold towards worldly interests and, instead, turned to mysticism.

Sari al-Saqati was in conversation with famous Sufis of the period such as Maruf Karkhi, Harith al-Muhasibi, and Bishr Hafi, and is the uncle and master of Junayd of Baghdad.

Sari al-Saqati's tales, words and ideas were generally transmitted by Junayd.

During his travels from Baghdad to the northern regions, Sari al-Saqati had the opportunity to meet many Sufis. He entered a zawiyah in Abadan, which belonged to the Sufis of Basra. Ali al-Jurjani, whom he met during his journey, advised him to go to Syria. In Syria, he was influenced by the Sufis who continued Ibrahim bin Adham's understanding of mysticism based on futuwwa and sincerity. He lived in Damascus, Ramla, Jerusalem and Tarsus; and while there, he participated in the jihad against the Byzantines. When he was in his sixties, he settled in Baghdad in 218 (833), and lived there until the end of his life. The grave of Sari al-Saqati is next to Junaid al-Baghdadi's in the Shunuziyya Cemetery in Baghdad.

62. Harith al-Muhasibi

Abu Abdullah Harith bin Asad bin Abdullah al-Anizi al-Basri

Al-Muḥāsibī for short (Arabic: المحاسبي),

(born in Basra in 781 AD – 857),

was the founder of the Baghdad School of Islamic philosophy, and a teacher of the Sufi masters Junayd al-Baghdadi and Sirri Saqti. He hailed from the Arab Anazzah tribe.

Scientific Contributions

Muhasibi means self-inspection/audit. It was his characteristic quality. He was a founder of Sufi doctrine, and influenced many subsequent Mukallemun, such as al-Ghazali.

Al-Muḥāsibī is the author of approximately 200 works; he wrote about theology and Tasawwuf (Sufism).

His father became wealthy, though al-Muhasibi refused wealth. Despite the affluent lifestyle available to him, true to his name "Al-Muḥāsibī", he retained quality of zuhd from Al-Hasan al-Basri.

The Sufis of his time had taken on certain practices, such as wearing woolen clothing, reciting the Qur'an at night, and limiting the kind and quantity of food eaten.

Al-Muḥāsibī observed that Sufi practices can help control the passions; but he also observed that they can result in problems like hypocrisy and pride. When outward piety becomes a part of one's image, it can mask hidden problems with the ego.

Al-Muḥāsibī emphasized that both the inner and outward states must be rectified. His proposed method for developing awareness of the inner self and purifying the heart was "Al-muhasabah" (constant self-examination) in anticipation of the Day of Judgement. Hence his name "Al-Muḥāsibī".

Among his works are: Kitab al-Khalwa and Kitab al-Ri`aya li-huquq Allah. In Kitab al-Khalwa, a discourse on fear and hope, he states the following.

> Know that the first thing that corrects you and helps you correct others is zuhd to detach from this world. It is attained by realization. Consideration is attained by reflection. If you think of this world, you will not find it worth sacrificing your ruh and iman for it. But you will find your ruh worthier of honor by ridiculing this world. This world is abhorred of God almighty and the messengers. It is an abode of affliction and a station of foolishness. Be on your guard from it.

Biographical Summary

Al-Muhasibi was born in 851 AD in Basra. The parents of Al-Muḥāsibī left Basra for Baghdad shortly after his birth, perhaps inclined to the economic opportunities in the new capital.

Al-Muhasibi later joined a group of scholars of kalam, led by Abdullah ibn Kullāb (died 855). They criticized the Jahmis, Mu'tazilis, and Anthropomorphists. The Mu'tazilis argued that the Qur'an was created, while Ibn Kullab argued against the createdness of the Qur'an by introducing a distinction between the speech of God (kalam Allah) and its realization: God is eternally speaking (mutakallim), but he can only be mukallim, addressing Himself to somebody, if this addressee exists.

In 848 (or possibly 851), the caliph al-Mutawakkil ended the Mihna, and, two years later, banned the Mu'tazilites' theology.

Al-Muhasibi died in 857 AD.

63. ansour al-Hallaj

Mansour al-Hallaj

(Arabic: ابو المغيث الحسين بن منصور الحلاج)

(c. 858 – 26 March 922) (Hijri c. 244 AH – 309 AH)

was a man of ruhaniyyat, a poet, and a Sufi. He is best known for his saying, "I am the Truth" ("*Ana'l-Ḥaqq*"). Al-Hallaj gained a wide following as a preacher before he became implicated in power struggles of the Abbasid court and was executed after 9 years of confinement on religious and political charges.

Scientific Contributions

In Mecca he made a vow to remain for one year in the courtyard of the sanctuary in fasting and total silence. When he returned from Mecca, he laid down the Sufi tunic and adopted a non-sufi lay style in order to be able to preach more freely. At the time, some Muʿtazilis and Shias who held high posts in the government accused him of deception and incited the mob against him. Al-Hallaj left for eastern Iran and remained there for five years, preaching fortified monasteries that housed volunteer fighters in the jihad. After that, he was able to return and install his family in Baghdad.

Al-Hallaj made his second pilgrimage to Mecca with four hundred disciples, where some Sufis, his former friends, accused him of sorcery and making a pact with the jinn. Afterwards he set out on a long voyage that took him to India and Turkestan. About 290/902 he returned to Mecca for his final pilgrimage clad in an Indian loin-cloth and a patched garment over his shoulders.

> There he prayed to God to be made despised and rejected, so that God alone might grant grace to Himself through His servant's lips.

When al-Hallaj returned to Baghdad from his last pilgrimage to Mecca, he built a model of the Kaaba in his home for private worship. Also, al-Hallaj began making proclamations that aroused popular emotion and caused anxiety among the educated classes. These included avowing his burning love for God and his desire to "die accursed by the Community", and statements such as "O Muslims, save me from God" ... "God has made my blood lawful to you: kill me". It was at that time that al-Hallaj is said to have pronounced: "I am the Truth".

He was denounced at the court, but a Shafi'i jurist refused to condemn him, stating that spiritual inspiration was beyond his jurisdiction and his words were not proof of disbelief. Al-Hallaj was spared execution, but he was punished for talking about being at one with God by being shaved, pilloried and beaten with the flat of a sword."

Al-Hallaj's preaching had by now inspired a movement for moral and political reform in Baghdad. In 296/908 Sunni reformers made an unsuccessful attempt to depose the underage caliph al-Muqtadir. When he was restored, his Shi'i vizier unleashed anti-Hanbali repressions which prompted al-Hallaj to flee Baghdad, but three years later he was arrested, brought back, and put in prison, where he remained for nine years.

The conditions of al-Hallaj's confinement varied depending on the relative sway his opponents and supporters held at the court, but he was finally condemned to death in 922 on the charge of being a Qarmatian rebel who wished to destroy the Kaaba, because he had said "the important thing is to proceed seven times around the Kaaba of one's heart."

According to another report, the pretext was his recommendation to build local replicas of the Kaaba for those who are unable to make the pilgrimage to Mecca. The queen-mother interceded with the caliph who initially revoked the execution order, but the intrigues of the vizier finally moved him to approve it. On 23 Dhu 'l-Qa'da (25 March) trumpets announced his execution the next day. The words he spoke during the last night in his cell are collected in *Akhbar al-Hallaj.* Thousands of people witnessed his execution on the banks of the Tigris River. He was first punched in the face by his executioner, then lashed until unconscious, and then decapitated or hanged. Witnesses reported that al-Hallaj's last words under torture were:

> "all that matters for the zahid is that the *Unique should reduce him to Unity*",

after which he recited the Quranic verse 42:18:

> "Those who disbelieve in it ⌜ask to⌝ hasten it ⌜mockingly⌝. But the believers are fearful of it, knowing that it is the truth. Surely those who dispute about the Hour have gone far astray."

His body was doused in oil and set alight, and his ashes were then scattered into the river. A cenotaph was "quickly" built on the site of his execution, and "drew pilgrims for a millennium"[16] until being swept away by a Tigris flood during the 1920s.

Some people question whether al-Hallaj was executed for religious reasons as has been commonly assumed. Sadakat Kadri points out that "it was far from conventional to punish heresy in the tenth century," and it is thought he would have been spared execution except that the vizier of caliph al-Muqtadir wished to discredit "certain figures who had associated themselves" with al-Hallaj.

Al-Hallaj addressed himself to popular audiences encouraging them to find God inside their own souls, which earned him the title of "the carder of innermost souls" (Hallāj al-asrār). He preached without the traditional Sufi habit and used language familiar to the local Shi'i population. This may have given the impression that he was a Qarmatian missionary rather than a Sufi.

Al-Hallaj was popularly credited with numerous karamat. He was said to have "lit four hundred oil lamps in Jerusalem's Church of the Holy Sepulchre with his finger and extinguished an eternal flame in a Zoroastrian fire temple with the tug of a sleeve."

Among other Sufis, al-Hallaj was an anomaly. Many Sufi masters felt that it was inappropriate to share mysticism with the masses, yet al-Hallaj openly did so in his writings and through his teachings. This was exacerbated by occasions when he would fall into trances which he attributed to being in the presence of God.

Hallaj was also accused of ḥulūl, the basis of which charge seems to be a disputed verse in which is proclaimed mystical union in terms of two spirits in one body. This position was criticized for not affirming union and unity strongly enough.

There are conflicting reports about his most famous statement, " أنا الحق *Anā l-Ḥaqq*" meaning "I am The Truth" which was taken to mean that he was claiming to be God, since *al-Ḥaqq* "the Truth" is one of the names of God. While meditating, he uttered انا الحق. The earliest report, coming from a hostile account of Basra grammarians, states that he said it in the mosque of al-Mansur, while testimonies that emerged decades later claimed that it was said in private during consultations with Junayd Baghdadi. Even though this utterance has become inseparably associated with his execution in the popular imagination, owing in part to its inclusion in his biography by Attar of Nishapur, the historical issues surrounding his execution are far more complex. In another controversial statement, al-Hallaj claimed "There is nothing wrapped in my turban but God," and similarly he would point to his cloak and say, ما في جبتي إلا الله *Mā fī jubbatī illā l-Lāh* "There is nothing in my cloak but God." He also wrote:

> I saw my Lord with the eye of the heart
>
> I asked, 'Who are You?'
>
> He replied, 'You'.

In the 11th volume of the proto-Salafist Ibn Kathir's book *al-Bidaya wa-l-Nihaya*, it is said that al-Hallaj used to deceive people by putting on plays with his hired men under the guise of spiritual healing, and extorting money from them by cunning and secret, and it is also stated that, he came to India to learn and practice Indian magic. Ibn Kathir also said in the book, "Abu Abd al-Rahman al-Sulami Amr ibn Uthman said on the authority of al-Makki: He said: "I was walking with al-Hallaj in some streets of Makkah and I read the Qur'an. I was reciting, and he heard my recitation and said: I can recite the same (recitation), so I left him". Narrated by Ibn Kathir, Abu Zari al-Tabari said, I heard Abu Ya'qub al-Aqta say: I gave my daughter in marriage to al-Husayn al-Hallaj when I saw his good conduct and diligence, and after a short time it became clear to me that He is a deceitful sorcerer, a hateful infidel.[26] Ibn Kathir also said, "Muhammad ibn Yahya al-Razi said: I heard Amr ibn Uthman cursing him and saying: If I could have killed him, I would have killed him with my own hands. I said to him: What did the Shaykh get on him? He Said: "I read a verse of the Book of Allah and He said: I can compose like it and speak like it." Ibn Kathir also said, and Abu al-Qasim al-Qushayri mentioned in his letter in the chapter on preserving the hearts of the sheikhs: Amr bin Uthman entered the house of al-Hallaj when he was in Makkah, he (Hallaj) was writing something on paper and he (Amr) said to him: What is it? He (Hallaj) said: It is *against the Qur'an.* He said: Then he prayed for him and then he was not successful. Hallaj denied that Abu Ya'qub al-Aqta married him to his daughter.

Al-Hallaj's principal works, all written in Arabic, included:

- Twenty-seven *Riwāyāt* (stories or narratives) collected by his disciples in about 290/902.
- Poems collected in *Dīwān al-Hallāj.*
- Pronouncements including those of his last night collected in *Akhbār al-Hallāj.*
- His best-known written work is the *Kitāb al-Tawāsīn*, (كتاب الطواسين), a series of eleven short works. In this book he used line diagrams and symbols to help him convey mystical experiences that he could not express in words.

Ṭawāsīn is the broken plural of the word *ṭā-sīn* which spells out the letters ṭā (ط) and sīn (س) placed for unknown reasons at the start of some surahs in the Quran. The chapters vary in length and subject. Chapter 1 is an homage to Muhammad, for example, while Chapters 4 and 5 are treatments of his legendary ascent to Mi'raj. Chapter 6 is the longest of the chapters and is devoted to a dialogue of Satan (Iblis) and God, where Satan refuses to bow to Adam, although God asks him to do so. Satan's monotheistic claim - that he refused to bow before any other than God even at the risk of eternal rejection and torment - is combined with the lyrical language of the love-mad lover from the Majnun tradition, the lover whose loyalty is so total that there is no path for him to any "other than" the beloved. This passage explores the issues of mystical knowledge (ma'rifa) when it contradicts God's commands for *although Iblis was disobeying God's commands, he was following God's will.* His refusal is due, others argue, to a misconceived idea of God's uniqueness and because of his refusal to abandon himself to God in love. Al-Hallaj stated in this book:

If you do not recognize God, at least recognize His sign; I am the creative truth because through the truth, I am eternal truth.

-Al-Hallaj, Kitāb al-Tawāsīn

Few figures in Islam provoked as much debate among classical commentators as al-Hallaj. The controversy cut across doctrinal categories. In virtually every major current of juridical and ilm-al-kalam thought (Hanafi, Maliki, Shafi'i Hanbali, Maturidi, Ash'ari, and also Shia Jafari) one finds his detractors and others who accepted his legacy completely or justified his statements with some excuse. His admirers among philosophers included Ibn Tufayl, Suhrawardi, and Mulla Sadra.

Although the majority of early Sufi teachers condemned him, he was almost unanimously canonized by later generations of Sufis. The principal Sufi interpretation of the *shathiyat* which took the form of "I am" sayings contrasted the permanence (*baqā*) of God with the mystical annihilation (*fanā*) of the individual, which made it possible for God to speak through the individual.

Some Sufi authors claimed that such utterances were misquotations or attributed them to immaturity, madness or intoxication, while others regarded them as authentic expressions of spiritual states, even profoundest experience of divine realities, which should not be manifested to the unworthy. Some of them, including al-Ghazali, showed ambivalence about their apparently blasphemous nature while admiring the spiritual status of their authors. Rumi wrote: "When the pen (of authority) is in the hand of a traitor, unquestionably Mansur is on a gibbet." Perhaps so in fulfillment of al-Hallaj's prayer on his last pilgrimage.

The supporters of Mansur have interpreted his statement as meaning, "God has emptied me of everything but Himself. "According to them,

Mansur never denied God's oneness and was a strict monotheist. However, he believed that the actions of man, when performed in total accordance with God's pleasure, lead to a blissful unification with Him. Malayalam author Vaikom Muhammad Basheer draws parallel between "Anā al-Ḥaqq" and Aham Brahmasmi, the Upanishad Mahāvākya which means 'I am Brahman' (the Ultimate Reality in Hinduism). Basheer uses this term to intend God is found within one's self'. There was a belief among European historians that al-Hallaj was secretly a Christian, until the French scholar Louis Massignon presented his legacy in the context of Islamic mysticism in his four-volume work *La Passion de Husayn ibn Mansûr Hallâj.*

Hallaj is highly revered by Yezidis, perhaps in reference to the attitude of al-Hallaj in relation to Yazidi rebellion of Zanj against the Abbasids. Yazidis composed a few religious hymns devoted to al-Hallaj, elements of whose views as expressed in *Kitab al-Tawasin,* can be found in Yazidi religion.

Biographic Summary

Al-Hallaj was born around 858 in Pars Province of the Abbasid Empire to a cotton-carder (*Hallaj* means "cotton-carder") in an town called al-Bayḍā'. His father moved to a town in Wasit famous for its school of Quran reciters. Al-Hallaj studied Qur'an and became a hafiz before he was 12 years old. Later, he would often retreat from worldly pursuits to join other mystics in study at the school of Sahl al-Tustari. He was a Sunni Muslim.

When he was twenty, al-Hallaj moved to Basra, where he married and received his Sufi education from 'Amr Makkī. He got tired of the conflict that existed between his father-in-law and 'Amr Makkī, and he went to Baghdad. He studied sufi education from the famous Sufi teacher Junayd of Baghdad.

Through his brother-in-law, al-Hallaj found himself in contact with a Zaidi Shi'i clan that supported the Zanj Rebellion. When the Zanj rebellion was crushed, he set out on pilgrimage to Mecca to seek solace; though he went against the advice of Junayd.

64. Junaid al-Baghdadi

Abu-l-Qāsim al-Junayd ibn Muḥammad ibn al-Junayd al-Khazzāz al-Qawārīrī

(Arabic: أبو القاسم الجنيد بن محمد الخزاز القواريري),

(for short Junayd of Baghdad (Arabic: الجنيد البغدادي)),

(908 to 910 - 908 to 910 CE),

was a mystic and one of the most famous of the early sufis. He is a central figure in the ruhani lineage of many Sufi orders.

Scientific Contributions

Junayd taught in Baghdad throughout his lifetime and was an important figure in the development of Sufi habits. Like Hasan of Basra before him, he was widely revered by his students and disciples as well as quoted by other sufis. Because of his importance in Sufiism, Junayd was often referred to as the "Sultan". According to him:

> People need to "relinquish natural desires to wipe out human attributes; to discard selfish motives to cultivate spiritual qualities; to devote oneself to true knowledge to do what is best in the context of eternity; to wish good for the entire community to be truly faithful to God; and to follow the Prophet in the matters of the Shari'a." This starts with the practice of zuhd and continues with withdrawal from society, intensive concentration on ibadah and dhikr of God, ikhlas, and muraqaba respectively. Contemplation produces fana.

Junayd spent 40 years in his ruhani course praying while sacrificing his sleep and any other worldly desires, but then a conceit in his heart arose that he has achieved his goal. Then he was inspired by God that "*He who is not worthy of union, all his good works are but sins.*" The prayers which become a source of pride are useless, as true prayer makes a person more humble and devoted to God. His name became famous in many parts of the world despite the persecution he faced and the tongues of slander shot at him. Even then, he did not start preaching until 30 of the great saints indicated to him that he should now call men to God. However, he chose not to preach as yet, saying, "While the master is there, it is not seemly for the disciple to preach." After witnessing Muhammad in his dream commanding him to preach, he had to listen to Sirri Saqtiy. The intensity of ishq poured out of a speech of Junayd such that out of the 40 people he first preached, 18 died and 22 fainted. His caliph and most dear disciple was Abu Bakr Shibli.

Junayd helped establish the "sober" school of Sufi thought, which meant that he was very logical and scholarly about his definitions of various virtues, tawhid, etc.

> Sober Sufism is characterized by people who "experience fana, [and] do not subsist in that state of selfless absorption in God but find themselves returned to their senses by God. Such returnees from the experience of selflessness are thus reconstituted as renewed selves," just like an intoxicated person sobering up. For example, Junayd is quoted as saying, "The water takes on the color of the cup." While this might seem rather confusing at first, 'Abd al-Hakeem Carney explains it as: "When the water is understood here to refer to the Light of Divine self-disclosure, we are led to the important

concept of 'capacity,' whereby the Divine epiphany is received by the heart of any person according to that person's particular receptive capacity and will be 'colored' by that person's nature".

Biographical Summary

The exact birth date of Abu-l-Qāsim al-Junayd ibn Muḥammad ibn al-Junayd al-Khazzāz al-Qawārīrī (Arabic: أبو القاسم الجنيد بن محمد الخزاز القواريري) ranges from 210 to 215 AH according to Abdel-Kader. His death ranges between 296 to 298 AH (908 to 910 CE). It is believed that al-Junayd's ancestors come from Nihawand in modern-day Iran. Al-Junayd was raised by his uncle Sirri Saqti after being orphaned as a boy. Al-Junayd's early education included teachings from Abū Thawr, Abū 'Ubayd, al-Ḥārith al-Muḥāsibī, and Sarī ibn Mughallas.

Hagiography by Attar of Nishapur, in his book Tazkirat al-Awliya, says that he had felt the pain of separation from the divine since childhood. Regardless of spiritual sorrow, he was known for his quick understanding and discipline when Sirri Saqti accepted him as a disciple. According to Attar, Junayd was only seven years of age when Sirri Saqti took him along for the Hajj. In al-Masjid an-Nabawi, there were 400 sheikhs discussing the concept of 'thankfulness' whereby each expounded his own view. When Sirri Saqti told him to present his definition, Junayd said, "Thankfulness means that should not disobey God by means of the favor which He has bestowed upon you, nor make of His favor a source of disobedience." The sheikhs unanimously agreed that no other words could define the term better. Sirri Saqti asked Junayd from where he could learn all this. Junayd replied, "From sitting with you."

His traditional hagiography continues by stating that Junayd went back to Baghdad and took up selling glasses. However, he spent most of the time in prayer. He retired to the porch of Sirri Saqti's house and kept himself away from worldly matters, devoting his thoughts only to God.

65. Ibn Umail

Muḥammad ibn Umayl al-Tamīmī

(Arabic: محمد بن أميل التميمي),

(c. 900 to c. 960 AD),

was an Egyptian sufi who used the symbolism from chemistry.

Scientific Contributions

Ibn Umayl was a mystical and symbolic alchemist. He saw himself as following his "predecessors among the sages of Islam" in rejecting alchemists who take their subject literally. Although such experimenters discovered the sciences of metallurgy and chemistry, Ibn Umayl felt the symbolic meaning of alchemy is the precious sufi goal that is tragically overlooked.

He wrote:

"Eggs are only used as an analogy... the philosophers ... wrote many books on such things as eggs, hair, the biles, milk, semen, claws, salt, sulphur, iron, copper, silver, mercury, gold and all the various animals and plants ... But then people would copy and circulate these books according to the apparent meaning of these things, and waste their possessions and ruin their souls"

(The Pure Pearl chap. 1)

Here Ibn Umayl indirectly suggests that the conventional pursuit of Al-Chemy is a misguided reading of the symbolic writings of the earlier sages.

Moreover, Ibn Umayl wrote a Book of the Explanation of the Symbols, emphasizing that the sages spoke "a language in symbols" and that they "would not reveal it [the secret of the stone] except with symbols". In this book, he gives a huge list of names for the stone, the water, etc. thus referring to one inner mystery or religious experience, which - in contrast to an allegory - cannot be fully explained.

Ibn Umail presented himself as an interpreter of mysterious symbols. He set his treatise Silvery Water in an Egyptian temple Sidr wa-Abu Sîr, the Prison of Yasuf, where Joseph learned how to interpret the dreams of the Pharaoh. (Koran: 12 Yusuf)

"... none of those people who are famous for their wisdom could explain a word of what the philosophers said. In their books they only continue using the same terms that we find in the sages .... What is necessary, if I am a sage to whom secrets have been revealed, and if I have learned the symbolic meanings, is that I explain the mysteries of the sages."

Ibn Umails Book of the Ḥall ar-Rumūz (Solving the Riddles) can be considered as a summary of his Silvery Water and Starry Earth.

Following are some of the works of Ibn Umail.

- Ḥall ar-Rumūz (Solving the Riddles/Book of Explanation of the Symbols)
- ad-Durra an-Naqīya (The Pure Pearl)
- Kitāb al-Maghnisīya (The Book of Magnesia)
- Kitāb Mafātīḥ al-Ḥikma al-'Uẓmā (The Book of the Keys of the Greatest Wisdom)
- al-Mā' al-Waraqî wa'l-Arḍ an-Najmīya (The Silvery Water and the Starry Earth) that comprises a narrative; a poem Risālat ash-Shams ilā al-Hilâl (Epistola solis ad lunam crescentem, the letter of the Sun to the Crescent Moon)
- Al-Qasida Nuniya (Poem rhyming on the Letter Nun), with a commentary by Ibn Umail. Ms. Beşir Ağa (Istanbul) 505. For the poem without commentary see Stapelton's Three Arabic Treatises.
- Al-Qasida al-mīmīya (Poem rhyming on the Letter Mīm), with a commentary by Ibn Umail.

Following are some of the recent references to the works of Ibn Umayl.

- 12th century: al-Mā' al-Waraqī (Silvery Water) became a classic of Islamic Alchemy. It was translated into Latin in the twelfth or thirteenth century and was widely disseminated among alchemists in Europe often called Senioris Zadith tabula chymica (The Chemical Tables of Senior Zadith)
- 1339: In the al-Mâ' al-Waraqī transcript that is now in Topkapi Palace Library, Istanbul, the scribe added a note to the diagram that the sun represents the spirit (al-rūḥ) and the moon the soul (al-nafs) so the "Letter from the Sun to the Moon" is about perfecting the receptivity of ruh to nafas.
- 14th century: Chaucer's Canon's Yeoman's Tale has alchemy as a theme and cites Chimica Senioris Zadith Tabula (The Chemical Tables of Senior Zadith).
- 15th century: Aurora consurgens is a commentary by Pseudo Aquinas on a Latin translation of Al-mâ' al-waraqî (Silvery Water).
- 1605 Senioris Zadith filii Hamuelis tabula chymica (The Chemical Tables of Senior Zadith son of Hamuel) was printed as part I of Philosophiae Chymicae IV. Vetvstissima Scripta by Joannes Saur.
- 1660: The Chemical Tables of Senior Zadith retitled Senioris antiquissimi philosophi libellus was printed in volume 5 of the Theatrum chemicum.
- 1933 Three Arabic treatises on alchemy by Muhammad bin Umail (10th century AD), prints the three treatises in Arabic, and prints them in 13th century Latin as they were partially translated from the Arabic to Latin in 13th century. Printed in the journal Memoirs of the Asiatic Society of Bengal, Volume 12, Calcutta.
- 1997/2006: Corpus Alchemicum Arabicum 1A: An improved translation of Book of the Explanation of the Symbols. Kitāb Ḥall

ar-Rumūz with a commentary by the Jungian psychologist and scholar Marie-Louise von Franz

Ibn Umayl's works also contain an early commentary on the Emerald Tablet (a short and compact text attributed to Hermes Trismegistus) that is full of symbolism.

Biographical Summary

Very little is known about ibn Umayl. That is perhaps because he seems to have led an introverted life style, which he recommended to others in his writings. Statements in his writings, comparing the Alchemical oven with Egyptian temples suggest that he might have lived for some time in Akhmim, the former center of Alchemy. He also quoted alchemists that had lived in Egypt like Dhul-Nun al-Misri.

However, there are Western speculations about him.

In later European literature, ibn Umayl became known by a number of false names. For example, the translators made his title Sheikh into 'senior', and they made his honorific al-sadik into 'Zadith' and 'ibn Umail' becoming by erroneous translation 'filius Hamuel', 'ben Hamuel' or 'Hamuelis'. All these are false names for Ibn Umail and falsification happens because these Western writers feel no obligation to correctness, feel no need to refer back to the original sources, and merrily remain ignorant of Arabic civilization.

A Vatican Library catalogue lists one manuscript with the nisba al-Andalusī, suggesting a connection to Islamic Spain. However, it seems heuristic for Vatican to attribute it to Ibn Umail, because his writings suggest that he mostly lived and worked in Egypt, and also visited North Africa and Iraq (but not Andalusia). Even Vatican felt no need for verification and validation, went its merry way, discarding academic rigor.

This is what they did with the name of Ibn Umail. They did little better in translating the actual content of the book.

66. Al-Tawhīdī

ʿAlī ibn Muḥammad ibn ʿAbbās

(Arabic: علي بن محمد بن عباس), also known as Abū Ḥayyān al-Tawḥīdī (Arabic: أبو حيان التوحيدي),

(923–1023),

was an influential intellectual and thinker of the 10th century. Yāqūt al-Ḥamawī described him as "the philosopher of litterateurs and the litterateur of philosophers." However, he was neglected and ignored by the historians until Yāqūt wrote his book Muʿjam al-Udabāʾ (معجم الأدباء); it contained a biographical outline of at-Tawḥīdī, relying primarily on what al-Tawḥīdī had written about himself.

Scientific Contributions

Al-Tawḥīdī was highly critical of himself and unsatisfied with much of his work, and he burned many of his own books later in life. Nevertheless, he left a set of literary, philosophical, and Sufi works, which were distinctive in the history of the Arabic literature. Perhaps the most important works are:

- Al-Baṣā'ir wa al-Dhakhā'ir
- Al-Hawamil wa al-Shawamil
- Al-Imtāʿ wa al-Mu'ānasa, Book of Enjoyment and Bonhomie, is a collection of anecdotes and includes a chapter on zoology.
- Al-Isharat al-Ilahiyya
- Al-Muqabasat
- Al-Sadaqa wa al-Sadiq
- Mathalib al-Wazirain, Book on the Foibles of the Two Ministers, is a commentary on the political and cultural infighting of his day.

Bibliographical Summary

There are differing views on the dates of al-Tawḥīdī's birth and death. According to Tārīkh-i Sistān, he was born in 923 Near Baghdad or Fars. Al-Tawḥīdī had a difficult childhood. He was born into a poor family that sold dates called tawḥīd (hence his surname), and spent much of his childhood as an orphan in the care of his uncle, who treated him poorly.

After completing his studies, al-Tawḥīdī worked as a scribe for various parties in various cities in the Muslim world. His last known regular assignment was for Ebn Saʿdān, who he worked for from 980 until Saʿdān's execution in 985. During this time, he was a member of a literary circle centered around Abū Solaymān Manṭeqī Sejestānī, and most of what is known about the circle is through al-Tawḥīdī's work.

After Saʿdān's execution, al-Tawḥīdī doesn't appear to have had regular work as a scribe, although he continued to write. During his final twenty years of life, he lived in poverty and unrecognized. He is known to have been alive in 1009, and likely died in 1023 in Shiraz.

67. Abu al-Fazal Yemeni Tamimi

Abd al-Wāḥid b. ʿAbd al-ʿAzīz b. al-Ḥārith b. Asad al-Tamīmī or Abū al-Faḍl al-Tamīmī

(Arabic: ابوالفضل عبد الواحد تميمى),

(341–410 AH / 952–1020 CE),

was a 10th century saint who belonged to the Junaidia order. He was the son and disciple of Abu al-Hasan al-Tamimi. He was an ardent worshipper and ascetic.

Scientific Contributions

Among his most celebrated works is I'tiqad al-Imam al-Mubajjal Ahmad ibn Hanbal (also known as I'tiqad al-Imam al-Munabbal Abi 'Abd Allah Ahmad ibn Hanbal).

Other works of Abū al-Faḍl al-Tamīmī include the following.

- Khādim-ush-Sharī'ah (Guardian of the Sacred Law)
- Sālik-ut-Tarīqah (Wayfarer of the Ruhani Path)
- Wāqif-ul-Haqīqah (Unveiler of Divine Mysteries)

Abū al-Faḍl al-Tamīmī is often associated with Abu Bakr Shibli due to his character. This is probably because he gained beneficence from Abu Bakr Shibli although he took Bayat at the hands of his father Abdul Aziz bin Haris bin Asad al-Tamimi from whom he was given the Sufi khirqa.

Muhaddith Shah Waliullah Dehlawi is reported to have said, "Abū al-Faḍl al-Tamīmī wore the Khirqa from both 'Abdul Aziz al-Tamimi and Abu Bakr Shibli. This is reflected in many of the authentic chains of spiritual transmission."

Abū al-Faḍl al-Tamīmī spent most of his life guiding people often while travelling. Amongst his various disciples was his prominent khalifah, Mohammad Yousaf Abu-al-Farrah Turtoosi, also noted in various books.

Abu Al Fazal Abdul Wahid al-Tamimi's saintly lineage of Faqr was given to him through his father and Murshid Abdul Aziz bin Hars bin Asad al-Tamimi in the following order.

- Muhammad
- 'Alī bin Abī Ṭālib
- al-Ḥasan al-Baṣrī
- Habib al Ajami
- Dawud Tai
- Maruf Karkhi
- Sirri Saqti
- Junaid Baghdadi, the founder of Junaidia silsila
- Abu Bakr Shibli
- ʿAbd al-ʿAzīz b. al-Ḥārith b. Asad al-Tamimi
- Abū al-Faḍl al-Tamīmī

Abū al-Faḍl al-Tamīmī conferred his khilafat upon Mohammad Yousaf Abu al-Faraj Tarasusi and he continued the order.

Biographical Summary

Abū al-Faḍl al-Tamīmī was born in Hejaz in 842 AD. Not many details about his early life are known except that his family is from Yemen which is why he was often regarded as "Yemeni". His name was Abd al-Wāḥid and he was the son of Shaikh ʿAbd al-ʿAzīz b. al-Ḥārith b. Asad. 'Tamimi' was a part of his name as his family belonged to the tribe al-Tamimi of Arabia. He followed the Hanafi school of thought.

Abu Al Fazal Abdul Wahid Yemeni Tamimi died on 410 AH which is 1034 CE in Baghdad. He was buried in the mausoleum of Imam Ahmad b. Hanbal in Baghdad. This was during the Abbasid Caliphate.

68. Al-Qushayri

'Abd al-Karīm ibn Hūzān Abū al-Qāsim al-Qushayrī al-Naysābūrī

(Persian: عبدالکریم قُشَیری),

(Arabic: عبد الكريم بن هوازن بن عبد الملك بن طلحة أبو القاسم القشيري),

(986 - 1072 AD),

was a theologian known for his works on Sufism. He was born in 986 CE (376 AH) in Nishapur which is in Khorasan Province in Iran. This region was widely known as a center of Islamic civilization up to the 13th Century CE. He was the grandfather of the scholar Abd al-Ghafir al-Farsi, a contemporary of Al Ghazali.

Scientific Contributions

Laṭā'if al-Isharat bi-Tafsīr al-Qur'ān is a famous work of al-Qushayri that is a complete commentary of the Qur'an. He determined that there were four levels of meaning in the Qur'an. First, the Ibara which is the meaning of the text meant for the mass of believers. Second, the ishara, only available to the elite in Ruhaniyat and lying beyond the obvious verbal meaning. Third, laṭā'if, subtleties in the text that were meant particularly for saints. And finally, the ḥaqā'iq, which he said were only comprehensible to the prophets. This text placed him among the elite of the Sufi mystics and is widely used as a standard of Sufi thought.

His fame however, is due mostly to his Al-Risala al-Qushayriyya, or Al-Qushayrī's Epistle on Sufism. This text is essentially a reminder to the people of his era that Sufis had authentic ancestral tradition, as well as a defense of Sufism against the doubters that rose during that time of his life.

Al-Qushayri repeatedly acknowledges his debt to, and admiration for, his Sufi master throughout his Risala. Daqqaq was instrumental in introducing Qushayri to another outstanding Sufi authority of Khurasan,

Abu 'Abd al-Rahman al-Sulami (412/1021), who is quoted on almost every page of the Risala. It has sections where al-Qushayrī discusses the creed of the Sufis, mentions important and influential Sufis from the past, and establishes fundamentals of Sufi terminology, giving his own interpretation of those Sufi terms.

Al-Qushayrī finally goes through specific practices of Sufism and the techniques of those practices. This text has been used by many Sufi saints in later times as a standard, as is obvious from the many translations into numerous languages.

Biographical Summary

Al Qushayri was born in 986 AD into a privileged Arab family from among the Banu Qushayr who had settled near Nishapur. As a young man he received the education of a country squire of the time: adab, the Arabic language, chivalry and weaponry (istiʿmāl al-silāḥ). That all changed when he journeyed to the city of Nishapur and was introduced to the Sufi shaykh Abū ʿAlī al-Daqqāq. Daqqāq later became the master and teacher of the mystical ways to Qushayri.

Al Qushayri later married the daughter of Daqqāq, Fatima. After the death of Daqqāq, Qushayri became the successor of his master and father-in-law and became the leader of mystic assemblies in the madrasa that Abu Ali al-Daqqāq built in 1001 CE, which later became known as al-Madrasa al-Qushayriyya or "the Qushayri school". In later years Qushayri performed the pilgrimage in the company of Abū Muḥammad al-Juwaynī (d. 438/1047), the father of al-Juwayni, Imam al-Haramayn. He also traveled to Baghdad and the Hijaz.

During these travels Al Qushayri heard Hadith from various prominent Hadith scholars. Upon his return he began teaching Hadith, which is something he is famous for. He returned to Baghdad where the Caliph al-

Qa'im had him perform hadith teachings in his palace. After his return to Khurāsān, political unrest in the region between the Ḥanafī and Ashʿarī-Shāfiʿī factions in the city forced him to leave Nishapur, but he was eventually able to return and lived there until his death in 1072 AD, when the Seljuq vizier Nizam al-Mulk re-established the balance of power between the Ḥanafīs and the Shāfiʿīs.

He left behind six sons and several daughters between Fatima and his second wife and was buried near al-Madrasa al-Qushayriyya, next to his father in-law Abū ʿAlī al-Daqqāq.

69. Ali al-Hujwiri

Abu al-Hasan Ali ibn Uthman al-Jullabi al-Hujwiri

(Persian: ابو الحسن علی بن عثمان الجلابی الهجویری),

(known reverentially as Data Ganj Baksh),

(c. 1009-1072/77),

was an alim and sufi who authored *Kashf al-Mahjub*, an early treatise on Sufism. Born in the Ghaznavid Empire, al-Hujwiri is believed to have contributed "significantly" to the spread of Islam in South Asia through his preaching.

Scientific Contributions

Ali Hujwiri is famous for writing the renowned *Kashf al-maḥjūb* (*Unveiling of the Hidden*). The work presents itself as an introduction to the various aspects of Sufism and also provides biographies of the greatest sufis. The *Kashf al-maḥjūb* is the only extant work of Ali Hujwiri.

Egyptian Sufi alim Abul Azaem has translated this work into Arabic.

There are also the following unpreserved works:

- *Dīwān*, a collection of the Hujwiri's poems.
- *Minhāj al-Dīn* (*The Way of the Religion*), a work containing: two parts: a detailed account of those companions of Muhammad whom Ali Hujwiri deemed the precursors of the Sufis; and a full *biography* of the executed 10th-century mystic Mansur al-Hallaj (d. 922).
- *Asrār al-khiraq wa 'l-ma'ūnāt'*, a work on the woolen, patched garments worn by the Sufis of his time, and also awarded to the new initiates.
- An untitled work explaining the meaning behind the mystical sayings of Mansur al-Hallaj.

- *Kitāb al-bayān li-ahl al-'iyān*, a treatise on the orthodox interpretation of the Sufi ideal of Fana.
- *Kashf al-Asrār*, a short treatise on how to fully adopt the path of Tasawwuf, translated with in-depth commentary by El-Sheikh Syed Mubarik Ali Shah El-Gillani.

Regarding fiqh, Ali Hujwiri believed it was a spiritual necessity to follow one of the schools of fiqh, being himself a follower of the Hanafi fiqh. As such, Ali Hujwiri condemned as "heretics" all those who espoused ruhaniyyat without following all the precepts of the *sharīʿah*. He further denounced all those "who held that ... when the Truth is revealed the shari'ah is abolished". For Ali Hujwiri, all true and orthodox tasawwuf activities needed to take place within the full compliance with shari'ah.

Regarding dancing, Ali Hujwiri asserted that purely secular dancing "has no foundation either in the religious law of Islam or in the path of Sufism, because all reasonable men agree that it is a diversion when it is in earnest, and an impropriety when it is in jest". As such, he censured "all the traditions cited in its favour" as "worthless". As for the legitimate ecstatic experiences of some Sufis, whose bodies convulsed when their "heart [throbbed] with exhilaration and rapture" on account of their intense love of God, Ali Hujwiri declared that these movements only outwardly resembled dancing and opined that "those who call it 'dancing' are utterly wrong. It is a state that cannot be explained in words: 'without experience no knowledge'."

Regarding poetry, Ali Hujwiri deemed it lawful to listen to virtuous poetry, saying: "It is permissible to hear poetry. The Messenger heard it, and the Companions not only heard it but also spoke it." Due to these reasons, he censured those who "declare that it is unlawful to listen to any poetry whatever, and pass their lives in defaming their brother

Muslims." Regarding the hearing of secular poetry, however, Ali Hujwiri's opinion was far stricter, and he deemed it "unlawful" to hear poetry or love-songs that enticed the hearer to carnal desires through detailed descriptions "of the face and hair and mole of the beloved." In conclusion, he stated that those who regarded the hearing of such poetry "as absolutely lawful must also regard looking and touching as lawful, which is infidelity and heresy."

Regarding awliya, Ali Hujwiri supported the belief in their existence. As such, he stated: "You must know that the principle and foundation of Sufism and tasawwuf rests on awliya, the reality of which is unanimously affirmed by all the teachers, though every one has expressed himself in a different language." Elsewhere, he said: "God has awliya whom He has specially distinguished by His Friendship and whom He has chosen to be the governors of His Kingdom and has marked out to manifest by His Actions and has peculiarly favored with diverse kinds of karamaat and has purged of natural corruptions and has delivered from subjection to their lower soul and passion, so that all their thoughts are of Him and their intimacy is with Him alone. Such have been in past ages, and are now, and shall be hereafter until the Day of Resurrection, because God exalted this community above all others and has promised to preserve the religion of Muhammad. The visible proof of illumination is to be found among the awliya and the select of God."

Mu'in al-Din Chishti stayed at al-Hujwiri's mausoleum and offered a tribute to him:

گنج بخش فیضِ عالَم مظہرِ نورِ خدا ناقصاں را پیرِ کامل ، کاملاں را راہنما

Ganj Bakhsh-e-Faiz-e-Alam Mazhar-e-Nur-e-Khuda,

Na Qasaan-ra Pir-i Kamil, Kamilaan-ra Rahnuma.

It means that this granter of treasures, benefactor of the world, and the manifestation of the effulgence of God's illumination, is a perfect guide for the beginners and a guiding post for the perfect ones.

Bibliographic Summary

Ali Hujwiri was born in Ghazni, in present-day Afghanistan, in around 1009 to Uthman ibn Ali or Bu Ali. His father was a direct descendant of Imam Hasan ibn Ali. His genealogical chain supposedly goes back eight generations to Caliph Ali. According to the autobiographical information recorded in his own *Kashf al-maḥjūb*, it is evident that Ali Hujwiri travelled widely through the Ghaznavid Empire and beyond; spending considerable time in Baghdad, Nishapur, and Damascus, where he met many of the pre-eminent Ṣūfīs of his time. In matters of fiqh, he received training in the Hanafi law under various teachers. As for his Sufi training, his teacher was al-Khuttalī. For a short period, the sufi is believed to have lived in Iraq. His brief marriage during this period is said to have been unhappy. Eventually, Ali Hujwiri settled in Lahore, where he died with the reputation of a renowned preacher and teacher. After his death, Ali Hujwiri was unanimously regarded as a great sufi.

Ali Hujwiri's ruhani lineage is as follows: Habib al-Ajami; Dawud al-Ta'i; Maruf al-Karkhi; Sari al-Saqati; Junayd al-Baghdadi; Abu Bakr Shibli; Ali Husri Husri; AbulFazal Khutli; Ali al-Hujwiri.

Al-Hujwiri is believed to have contributed "significantly" to the acceptance of Islam in South Asia through his teaching. In the present day, al-Hujwiri is venerated as the main sufi of Lahore, Pakistan. He is, one of the most widely venerated sufis in the entire South Asia. His shrine in Lahore, popularly known as Data Darbar, is one of the most frequented shrines in South Asia. At present, it is Pakistan's largest shrine "in numbers of annual visitors and in the size of the shrine complex," and, having been nationalized in 1960, is managed today by the Department of Awqaf and Religious Affairs of the Punjab. The sufi himself remains a "household name" in South Asia.

70. Khwaja al-Ansari

Abu Ismaïl Abdullah al-Herawi al-Ansari or Abdullah Ansari of Herat

(Persian: خواجه عبدالله انصاری), (also known as Pir-i Herat ((پیر هرات) "Sage of Herat"),

(1006–1088),

was a Sufi saint who lived in the 11th century in Herat (modern-day Afghanistan). One of the outstanding figures of 11th century Khorasan, Ansari was a commentator of the Qur'an, scholar of the Hanbali school of thought (madhhab), Mohaddith, polemicist, and spiritual master, known for his oratory and poetic talents in Arabic and Persian.

Scientific Contributions

Khwaja Abd Allah al-Ansari excelled in the knowledge of Hadith, history, and ʻilm al-ansāb. He wrote several books on Islamic mysticism and philosophy in Persian and Arabic.

Khwaja Abd Allah al-Ansari was one of the first Sufis to write in Persian, which he wrote in a local dialect, thus indicating that he wanted to spread his teachings to the common people instead of the ulama, who knew Arabic.

His most famous work is "Munajat Namah" (literally 'Litanies or dialogues with God'), which is considered a masterpiece of Persian literature.

Following are some books (in Persian and Arabic) written by Khwaja Abd Allah al-Ansari.

- Munajat Namah (Persian: مناجات نامه)
- Nasayeh (Persian: نصایح)
- Zad-ul Arefeen (Arabic: زاد العارفين)
- Kanz-ul Salikeen (Persian: كنز السالكين)

- Haft Hesar (Persian: هفت حصار)
- Elahi Namah (Persian: الهی نامه)
- Muhabbat Namah (Persian: محبت نامه)
- Qalandar Namah (Persian: قلندر نامه)
- Resala-é Del o Jan (Persian: رساله دل و جان)
- Resala-é Waredat (Persian: رساله واردات)
- Sad Maidan (Persian: صد میدان)
- Resala Manaqib Imam Ahmad bin Hanbal (Arabic: رسالة مناقب الإمام أحمد بن حنبل)
- Anwar al-Tahqeeq (Arabic)
- Dhamm al-Kalaam (Arabic)
- Manāzel al-Sā'erīn (Arabic)
- Kitaab al-Frooq (Arabic)
- Kitaab al-Arba'een (Arabic)

Biographical Summary

Khwaja Abd Allah al-Ansari was born in the Kohandez, the old citadel of Herat, on 4 May 1006. His father, Abu Mansur, was a shopkeeper who had spent several years of his youth at Balkh.

Khwaja Abd Allah al-Ansari was a disciple of Abu al-Hassan al-Kharaqani. He practiced the Hanbali school of Sunni jurisprudence. The Shrine of Khwaja Abd Allah al-Ansari, built during the Timurid dynasty, is a popular pilgrimage site.

After his death, many of his sayings recorded in his written works transmitted by his students were included in the Tafsir of Maybudi, "Kashf al-Asrar" (The Unveiling of Secrets). This is among the earliest complete Sufi Tafsirs (Awalagyy) of the Quran and has been published several times in 10 volumes.

He used to avoid the company of the rich, powerful and the influential. His yearly majlis-e wa'az was attended by people from far and wide. Whatever his disciples and followers used to present to him was handed over to the poor and the needy. He is said to have had a very impressive personality, and used to dress gracefully.

Khwajah Abdullah Ansari of Herat was a direct descendant of Abu Ayyub al-Ansari, a companion of the Islamic prophet Muhammad, being the ninth in line from him. The lineage is described, and traced in the family history records, as follows;

Abu Ismail Khajeh Abdollah Ansari, son of Abu Mansoor Balkhi, son of Jaafar, son of Abu Mu'aaz, son of Muhammad, son of Ahmad, son of Jaafar, son of Abu Mansoor al-Taabi'i, son of Abu Ayyub al-Ansari.

In the reign of the third Rashid Caliph, Uthman, Abu Mansoor al-Taabi'i took part in the conquest of Khorasan, and subsequently settled in Herat, his descendant Khwajah Abdullah Ansari died there in 1088.

The Hanbali jurist ibn Qayyim al-Jawziyya wrote a lengthy commentary on a treatise written by Khwajah Abdullah Ansari entitled “Madarij al-Salikin”. He expressed his love and appreciation for Khwajah Abdullah Ansari in this commentary with his statement, "Certainly I love the Sheikh, but I love the truth more!". Ibn Qayyim al-Jawziyya refers to Khwajah Abdullah Ansari with the honorific title "Sheikh al-Islam" in his work “Al-Wabil al-Sayyib min al-Kalim al-Tayyab”.

71. Abu Saeed Mubarak Makhzoomi

Abu Saeed Mubarak Makhzoomi

(Arabic: ابوسعیدمبارک مخزومی), known also as Mubarak bin Ali Makhzoomi

and Abu Sa'd al-Mubarak (occasionally known as Qazi Abu Sa'd al-Mubarak al-Mukharrimi),

(1013 – 1119 AD),

was a Sufi saint as well as theologian and a Hanbali jurist. He was based in Baghdad, Iraq. Abu Saeed was the Murshid and spiritual guide of Shaikh Abdul Qadhir Jilani.

Scientific Contributions

Abu Saeed Mubarak Makhzoomi was a renowned Imam of Fiqh in his era. He followed the Hanbali school of thought.

He was the Murshid and spiritual guide of Shaikh Abdul Qadhir Jilani. He often said: "I invested Shaikh Abdul Qadhir Jilani with a Khirqa and he invested me too with a Khirka. We attained blessings from each other."

Biographical Summary

Abu Saeed Mubarak Makhzoomi was born in Hankar (the land of his Murshid) on 12th Rajab 403 Hijri (1013 AD) but spent most of his life in Makhzum, a small town in Baghdad. He established Baab-ul-Azj, the famous madrasa of Baghdad whom he later handed over to his disciple and khalifah, Shaikh Abdul Qadhir Jilani. Abu Saeed Mubarak Makhzoomi was also appointed as the chief justice but he preferred to renounce the worldly life. Thereafter he led his life as a mystic and devoted his time to the dhikr of Allah. He died on 11th Rabī' al-Thānī 513 Hijri (1119 AD) and was buried in Baab-ul-Azj, Baghdad.

The lineage of Faqr reaches Abu Saeed Mubarak Makhzoomi from Muhammad in the following order:

- Mohammad
- 'Alī bin Abī Ṭālib
- al-Ḥasan al-Baṣrī
- Habib al Ajami
- Dawud Tai
- Maruf Karkhi
- Sirri Saqti
- Junaid Baghdadi, founder of the Junaidia order
- Abu Bakr Shibli
- Abdul Aziz bin Hars bin Asad Yemeni Tamimi
- Abu Al Fazal Abdul Wahid Yemeni Tamimi
- Mohammad Yousaf Abu al-Farah Tartusi
- Abu-al-Hassan Ali Bin Mohammad Qureshi Hankari
- Abu Saeed Mubarak Makhzoomi

Abu Saeed Mubarak Makhzoomi conferred khilafat upon Shaikh Abdul Qadhir Jilani who continued the order as Qadri order.

72. Abdul Qadhir Gilani

Abdul Qadhir Gilani

(Arabic: عبد القادر الجيلاني),

(1078 - 1166 AD in Baghdad),

was a Hanbali and Sha'fi faqih, preacher, and Sufi who was the eponym of the Qadhiriyya Tariqa, one of the oldest.

Scientific Contributions

In 1095, he went to Baghdad. There, he pursued the study of Hanbali law under Abu Saeed Mubarak Makhzoomi and ibn Aqil. He studied hadith with Abu Muhammad Ja'far al-Sarraj. His Sufi spiritual instructor was Abu'l-Khair Hammad ibn Muslim al-Dabbas.

After completing his education, Gilani left Baghdad. He spent twenty-five years wandering in the deserts of Iraq.

Gilani belonged to the Shafi'i and Hanbali schools of fiqh. He placed Shafi'i fiqh on an equal footing with the Hanbali fiqh, and used to give fatwa according to both of them simultaneously. This is why al-Nawawi praised him in his book entitled Bustan al-'Arifin (Garden of the Sufis), saying: We have never known anyone more dignified than Baghdad's Sheikh Muhyi al-Din 'Abd al-Qadhir al-Gilani, may Allah be pleased with him, the Sheikh of Shafi'is and Hanbalis in Baghdad.

In 1127, Gilani returned to Baghdad and began to preach to the public. He joined the teaching staff of the school belonging to his teacher, al-Mazkhzoomi, and was popular with students. In the morning, he taught hadith and tafsir, and in the afternoon he discoursed on the science of the heart and the virtues of the Quran. He was said to have been a convincing preacher who converted numerous Jews and Christians and *who integrated Sufi mysticism with fiqh.*

Abdul Qadhir Jilani converted thousands of people to Islam through his compassionate and inclusive approach to Inner purification and devotion towards Allah. His emphasis on inner purification, divine love, and ethical living resonated deeply with many, attracting followers from diverse backgrounds.

One of Shaykh Abdul Qadhir Jilani's most significant contributions was the establishment of the Madrasah al-Qadhiriyya in Baghdad. This institution became a center for learning and tasawwuf, attracting students from various regions. The curriculum included the study of the Qur'an, Hadith, Fiqh and Tasawwuf, providing a comprehensive religious education.

The influence of Shaykh Abdul Qadhir Jilani extended to political and military leaders of his time. His teachings inspired rulers to adopt more just and ethical governance. Prominent figures such as Nur ad-Din Zangi and Salahuddin Ayyubi were known to respect and follow the principles advocated by the Shaykh, which contributed to their own reforms and successes.

Abdul Qadhir Jilani wrote many books.

- Kitab Sirr al-Asrar wa Mazhar al-Anwar (The Book of the Secret of Secrets and the Manifestation of Light)
- Futuh al ghaib (Secrets of the Unseen)
- Jila' al-Khatir (The Purification of heart)
- Ghunya- tuṭ-ṭalibeen) (Treasure for Seekers) غنيه الطالبين
- Al-Fuyudat al-Rabbaniya (Emanations of Lordly Grace)
- Khamsata 'Ashara Maktuban (Fifteen Letters)
- Kibriyat e Ahmar
- A Concise Description of Jannah & Jahannam
- The Sublime Revelation (al-Fatḥ ar-Rabbānī)

Biographical Summary

Abdul Qadhir Gilani was born in 1077 or 1078 in the town of Na'if, Rezvanshahr in Gilan, Persia, and died in 1166 in Baghdad.

He is also known as Gauth Al-Azam.

The honorific Muhiyudin denotes his status with many Sufis as a "reviver of religion".

Gilani refers to his place of birth, Gilan, which is really is the name of a province in Iran rather than a place or city in that province. According to the book Al-Nujūm al-Zāhira by the 15th-century historian Ibn Taghribirdi (died 1470), Abdul Qadhir was born in Jil in Iraq. So he is also known as Abdul Qadhir Jilani. In Persian and Urdu his nisbah is written both with a "jeem" and a "ghaaf".

Al-Gilani died in 1166 and was buried in Baghdad.

During the reign of the Safavid Shah Ismail I, Gilani's shrine was destroyed. However, in 1535, the Ottoman emperor Suleiman the Magnificent had a dome built over the shrine.

73. Shihab al-Din Yahya ibn Habash Suhrawardi

Shihāb ad-Dīn" Yahya ibn Habash Suhrawardī

(for short Suhrawardī),

(Persian: شهاب‌الدین سهروردی),

(1154–1191 AD),

was a Sufi philosopher who founded Suhrawardīa sufi tariqa. He wrote *Kitab Hikmat al-ishraq* "Wisdom in Divine Illuminations", inspired by the Verse of Light in the Holy Quran. (it is inappropriately translated as Philosophy of Illuminism).

Scientific Contributions

It is apparent that Suhrawardi was strongly influenced by the Sufi tradition, bearing in mind his journeys that he considered a necessary prerequisite.

Like al-Hallaj Suhrawardi created a symbolic language to give expression to his wisdom (*hikma*). Contemplating on the "Verse of Light" in the Holy Quran, Suhrawardi taught a complex and profound cosmology, in which all creation is a successive outflow from the original Supreme Divine Light (*Nurun ala Nur*). The fundamental of his wisdom is pure immaterial light, where nothing is manifest, and which unfolds from the Divine Light in a descending order of ever-diminishing intensity and, through complex interaction, gives rise to a an array of lights and shadows. In other words, the universe and all levels of existence are but varying degrees of light and darkness, using a taxonomy of bodies, categorizing objects in terms of their reception or non-reception of light.

In accord with Quranic statement, in the realm of eternity, "Alastu birabbikum" "Am I not your Rab?" Suhrawardi considers a previous existence for every soul in the angelic realm before its descent to the realm of the body. The soul is divided into two parts, one remains in heaven and the other descends into the dungeon of the body. The human soul is always sad because it has been separated from its source. Therefore, it aspires to become reunited with it. The soul can only reach felicity again when it is united with its source. Suhrawardi holds that the soul should seek felicity by detaching itself from its tenebrous body and worldly matters and access the world of immaterial light. The souls of awliya, after leaving the body, ascend even above the angelic world to enjoy proximity to the Divine Light, which is the only absolute Reality.

According to Hossein Nasr since Sheykh Ishraq was not translated into Western languages in the medieval period, Europeans had little knowledge about Suhrawardi and his philosophy. His school is ignored even now by later scholars. Sheykh Ishraq tried to pose a new perspective on questions like the question of Existence. He not only caused peripatetic philosophers to confront new questions but also gave new life to the body of philosophy after Avicenna.

Suhrawardi left over 50 writings in Persian and Arabic. Below are included modern editions and translations, substantially edited.

- *Partaw Nama* ("Treatise on Illumination")
- *Hayakal al-Nur* al-Suhrawardi [Sohravardi, Shihaboddin Yahya] (1154–91)
- *Hayakil al-nur* ("The Temples of Light"), ed. M.A. Abu Rayyan, Cairo: al-Maktaba al-Tijariyyah al-Kubra, 1957. (The Persian version appears in oeuvres vol. III.)
- *Alwah-i Imadi* ("The tablets dedicated to Imad al-Din")
- *Lughat-i Muran* ("The language of Termites")
- *Risalat al-Tayr* ("The Treatise of the Bird")
- *Safir-i Simurgh* ("The Calling of the Simurgh")

- *Ruzi ba Jama'at Sufiyaan* ("A Day with the Community of Sufis")
- *Fi Halat al-Tufulliyah* ("On the State of Childhood")
- *Awaz-i Par-i Jebrail* ("The Chant of Gabriel's Wing")
- *Aql-i Surkh* ("The Red Intellect")
- *Fi Haqiqat al-'Ishaq* ("On the Reality of Love")
- *Bustan al-Qolub* ("The Garden of Hearts")

- *Kitab al-talwihat*
- *Kitab al-moqawamat*
- *Kitab al-mashari' wa'l-motarahat*, Arabic texts edited with introduction in French by H. Corbin, Tehran: Imperial Iranian Academy of Philosophy, and Paris: Adrien Maisonneuve, 1976; vol II: I. *Le Livre de la Théosophie oriental*
- *Kitab Hikmat al-ishraq* 2. *Le Symbole de foi des philosophes.* 3. *Le Récit de l'Exil occidental*, Arabic texts edited with introduction in French by H. Corbin, Tehran: Imperial Iranian Academy of Philosophy, and Paris: Adrien Maisonneuve, 1977; vol III: *oeuvres en persan*, Persian texts edited with introduction in Persian by S.H. Nasr, introduction in French by H. Corbin, Tehran: Imperial Iranian Academy of Philosophy, and Paris: Adrien Maisonneuve, 1977. (Only the metaphysics of the three texts in Vol. I were published.) Vol. III contains a Persian version of the *Hayakil al-nur*, ed. and trans. H. Corbin
- *L'Archange empourpré: quinze traités et récits mystiques*, Paris: Fayard, 1976, contains translations of most of the texts in vol. III of *oeuvres philosophiques et mystiques*, plus four others. Corbin provides introductions to each treatise, and includes several extracts from commentaries on the texts. W.M. Thackston, Jr, *The Mystical*

and Visionary Treatises of Shihabuddin Yahya Suhrawardi, London: Octagon Press, 1982, provides an English translation of most of the treatises in vol. III of *oeuvres philosophiques et mystiques*, which eschews all but the most basic annotation; it is therefore less useful than Corbin's translation from a philosophical point of view)

- *Mantiq al-talwihat*, ed. A.A. Fayyaz, Tehran: Tehran University Press, 1955. The logic of the Kitab al-talwihat (The Intimations)
- *Kitab hikmat al-ishraq* (*The Philosophy of Illumination*), trans H. Corbin, ed. and intro. C. Jambet, *Le livre de la sagesse orientale: Kitab Hikmat al-Ishraq*, Lagrasse: Verdier, 1986. (Corbin's translation of the Prologue and the Second Part (The Divine Lights), together with the introduction of Shams al-Din al-Shahrazuri and liberal extracts from the commentaries of Qutb al-Din al-Shirazi and Mulla Sadra. Published after Corbin's death, this copiously annotated translation gives to the reader without Arabic immediate access to al-Suhrawardi's illuminationist method and language)
- *The Philosophy of Illumination: A New Critical Edition of the Text of Hikmat Al-Ishraq*, edited by John Walbridge and Hossein Ziai, Provo, Brigham Young University Press, 1999.
- *The Shape of Light: Hayakal al-Nur*, interpreted by Shaykh Tosun Bayrak al-Jerrahi al-Halveti, Fons Vitae, 1998.
- *The Mystical & Visionary Treatises of Suhrawardi*, Translated by W.M. Thackson, Jr., London, The Octagon Press, 1982.

Biographical Summary

Suhrawardi was born in 1154 in Suhraward, a village located between the towns of Zanjan and Bijar Garrus in Iran. He learned wisdom and fiqh in the city of Maragheh. His teacher was Majd al-Dīn Jīlī who was also Fakhr al-Din al-Razi's teacher. He then went to Iraq and Syria for several years and developed his knowledge while he was there.

His lived around forty years during which he produced a series of works that established him as the founder of a new Sufi tariqa, called Suhrawardia. In 1186, at the age of thirty-two, he completed his magnum opus, *hikmat al-Ishraq* ("Wisdom in Divine Illuminations"), inspired by the Verse of Light in the Holy Quran.

There are several reports of his death. The most commonly held view is that he was executed sometime between 1191 and 1208 in Aleppo on charges of cultivating Batini teachings and philosophy, by the order of al-Malik al-Zahir, son of Saladin. Other traditions hold that he starved himself to death, others tell that he was suffocated or thrown from the wall of the fortress, then burned by some people.

He is referred to by the honorific title *Shaikh al-ʿIshraq* "Master of Ishraq" and *Shaikh al-Maqtul* "the Master who was killed".

74. Abu Madyan

Abu Madyan Shuʿayb ibn al-Husayn al-Ansari al-Andalusi

(Arabic: ابو مدين شعيب بن الحسين الأنصاري الأندلسي), (commonly known as Abū Madyan),

(c. 1126 – 1198 CE),

was an influential Andalusian mystic and a great Sufi master.

Scientific Contributions

The basic principles and virtues taught at Madyan's school in Bejaia were repentance (tawba), asceticism (zuhd), paying visits to other masters, and service to experienced masters.

He emphasized futuwa (youth/chivalry) but only when accompanied by the obedience of devotees to their master, the avoidance of disagreements between devotees, justice, constancy, nobility of mind, the denunciation of the unjust, and a feeling of satisfaction with the gifts of God.

Abu Madyan and his followers refused to confine themselves to only asceticism and meditation alone, but instead lived day to day by maintaining close relationships with the people around them.

Along with sharing his knowledge and ideas with his disciples, Abu Madyan wrote many poems and spoke in proverbs in order to connect with the masses and not just the intellectuals.

According to Yahya B. Khaldun, Abu Madyan's teachings may all be summed up in this verse which he often repeated,

"Say Allah! and abandon all that is matter, or is connected with it, if though desirest to attain the truth goal."

Aside from attaining Ghawth status and teaching hundreds and hundreds of disciples, Abu Madyan left his mark in more ways than one. He gained immense popularity because he was relatable, despite his high scholarly status. He had a personality and way of speaking that united people from all walks of life, from the common people to the academics. Even to this day, scholars say that no one of the time surpassed him in religious and intellectual influence. His school produced hundreds of saints and out of the 46 Sufi saints in the Rif region, 15 were his disciples. People still visit his tomb today for asking god through him, they call it tawasul. They visit him from all around the world.

There are very few surviving writings from Abu Madyan, and of those that do still exist, there are mystical poems, a testament (wasiyya) and a creed (akida). He encouraged the free expression of emotions rather than rigidity, but also made known his support of asceticism, complete devotion to God, and a minimalist lifestyle.

Following are some of the works by Abu Madyan.

- Bidayat al Mouridin, Ms 938, Bibliot. Nat. Alger.
- Diwan, (collection of his poems) édit. Chaouar of Tlemcen, Damascus, 1938.
- Ouns al Wahid, Ms 2-105 (8) fol. 337–343, Bibliot. Nat. Paris, ed. in Cairo 1301–1884, with a commentary by Ahmed Bâ'chan.
- Tahfat al Arib, pub. et trad. in Latin par F. de Dombay, Vindobonae, Ebn Médirai Mauri Fessani Sentenciae quaedam arabicae, 1805
- The Way of Abu Madyan, bilingual collection, Islamic Texts Society, Cambridge, 1996. Transl by Vincent Cornell.
- Adab al-Murid, A poem on the etiquette of the murid for beginners on the spiritual path of suluk.

Biographical Summary

Abu Madyan was born in Cantillana, a small town about 35 km away from Seville, in 1126. He came from an obscure family and his parents were poor. As he grew up, he learned the trade of a weaver as it was a popular practice at the time. His insatiable hunger for knowledge, however, piqued his interest in the Qur'an and the study of religion and mysticism.

After crossing the Strait of Gibraltar, he worked for a while in Sabta (Ceuta) with fishermen. Afterwards, he went to Marrakesh, where he served in the Almoravid army defending the city.

Soon after, Abu Madyan traveled to Fez to complete his education. He left for Fez at about the end of the Almoravid era or at the beginning of the founding of the Almohad state. There, he studied under Abu Ya'azza al-Hazmiri, 'Ali Hirzihim, and al-Dakkak. It was al-Dakkak that provided him with the khirka, the cloak passed from Master to student in the study of Sufism.

Abu Madyan was particularly fascinated with mysticism by Sidi Ali Ibn Harazem. After finishing his studies with his master Abu Ya'za, he traveled to the Orient. During his time in the Orient, he became familiar with the works of Al-Ghazali, one of the most prominent theologians, philosophers, and mystics of Sunni Islam, regarded as one of the renewers of the religion.

Abu Madyan went to Mecca where he met the great Muslim saint, Jilani, and completed his spiritual training under him. On his return, he went to the town of Béjaïa where he practiced very strict asceticism and acquired an honorable reputation for his knowledge. People would come far to both listen to his public lectures and consult him on certain manners. People believed he could even perform miracles.

His beliefs were in opposition to the Almohad doctors of that town. The Almohads were disturbed at his increasing reputation and wanted to get rid of him.

Eventually, Abu Madyan settled in the town of Béjaïa where he established a mosque-school (zawiya). The sheer amount of fame and influence that Abu Madyan evoked raised serious concern from the political powers of the time. The Almohad Caliph Ya'qub al-Mansur summoned Abu Madyan to Marrakech for this reason so he could talk to Abu Madyan himself. Upon his summoning to Marrakech, Abu Madyan was taken ill and died before he reached his destination in 1198, near the river of Ysser (يسر).

His last sigh was supposedly "Allah al-Haqq." He was buried in al-'Ubbad near Tlemcen, Algeria. His funeral was widely commemorated by the people of Tlemcen and he has been considered the patron saint and protector of Tlemcen ever since. A mausoleum was built by the order of the Almohad sovereign, Muhammad al-Nasir, too shortly after his death. Many princes and kings of Tlemcen have contributed to this mausoleum since his demise. Many monuments, a good number of them still well preserved, were built in his honor next to his tomb by the Marinid kings, who controlled Tlemcen in the 14th century. One such monument is the Mosque of Madrasa.

His tomb became the center of fine architecture and is still a place of pilgrimage for many Sufis today.

We observe that the biography of Abu Madyan is instructive. During their lives, such people of Ruhaniyat are ordinary Muslims, often unnoticed. When they do become popular, often without themselves wanting it to happen, then many things can occur. These include, for example, but are not limited to, the following events: people start attributing miracles to them (without themselves necessarily claiming any); the rulers take note of their popularity and assess if they pose a danger to the rule; even if the rulers were opposed to them during their lifetime, the rulers want to mend fences with the popular sentiment, and a common practice is to please the public by building their mausoleum, and support charities (that the Sufi might or might not have approved); as the time passes the Sufi might get more and more enshrined, and further miracles and stories might flourish which might not have been noted during his or her lifetime.

In this light, one might think that as the Sufi approaches God, he sometimes also becomes approachable to the people. The people celebrate their own version of the Sufi; and this version is authored by the popular sentiment, often independently of the lifetime of the Sufi.

75. Farid ud-Din Attar Nishapuri

Abū Ḥāmid bin Abū Bakr Ibrāhīm

Persian: ابوحمید بن ابوبکر ابراهیم),

(better known as Farīd ud-Dīn (فریدالدین) and ʿAṭṭār Nishapuri (عطار نیشاپوری),

(c. 1145 – c. 1221),

was an Iranian poet, theoretician of Sufism, and hagiographer from who had an immense and an influence on Persian poetry and Sufism.

Scientific Contributions

Farid ud-Din Attar wrote a collection of lyrical poems and number of long poems in the philosophical tradition of Islamic tasawwuf, as well as a prose work with biographies and sayings of famous sufis. *The Conference of the Birds*, *Book of the Divine*, and *Memorial of the Saints* are among his best known works.

The thoughts depicted in Attar's works reflect the evolution of Sufiism. It starts with the idea that the release of the body-bound soul is awaited for the return to its source in the other world. It can be experienced during the present life in mystic union attainable through inward purification. In explaining his thoughts, Attar uses material from Sufi sources lake people of tasawwuf and zuhd. His talent for perception of deeper meanings behind outward appearances enables him to turn details of everyday life into illustrations of his thoughts. As a source on the phenomenology of Sufism, his works have immense value.

Attar's authorship of the *Dīwān* and the *Manṭiq-uṭ-Ṭayr* is well known; but there are other works such as *Mukhtār-Nāma* and *Khusraw-Nāma* and other titles mentioned in their prefaces. These include the following.

- Dīwān (دیوان)

- Asrār-Nāma (اسرارنامه)
- Manṭiq-uṭ-Ṭayr (منطق‌الطیر), also known as Maqāmāt-uṭ-Ṭuyūr (مقامات‌الطیور)
- Muṣībat-Nāma (مصیبت‌نامه)
- Ilāhī-Nāma (الهی‌نامه)
- Jawāhir-Nāma (جواهرنامه)
- Šarḥ al-Qalb (شرح‌القلب)

He states in the introduction of the *Mukhtār-Nāma*, that he himself destroyed the *Jawāhir-Nāma' and the* Šarḥ al-Qalb. One work is missing from these lists, namely the *Tadhkirat-ul-Awliyā*, which was probably omitted because it is a prose work. In its introduction Attar mentions three other works of his, including one entitled *Šarḥ al-Qalb*, presumably the same that he destroyed. The nature of the other two, entitled *Kašf al-Asrār* (کشف‌الاسرار) and *Ma'rifat al-Nafs* (معرفت‌النفس), remains unknown.

In **Manṭiq-uṭ-Ṭayr** *(The Conference of the Birds)*, the birds of the world gather to decide who is to be their sovereign, as they have none. The hoopoe, the wisest of them all, suggests that they should find the legendary Simorgh. The hoopoe leads the birds, each of whom represents a human fault which prevents human kind from attaining enlightenment. The hoopoe tells the birds that they have to cross seven valleys in order to reach the abode of Simorgh. These valleys are as follows:

- *1. Valley of the **Quest,*** where the Wayfarer begins by casting aside all dogma, belief, and unbelief.
- *2. Valley of **Love,*** where reason is abandoned for the sake of love.
- *3. Valley of **Knowledge,*** where worldly knowledge becomes utterly useless.
- *4. Valley of **Detachment,*** where all desires and attachments to the world are given up. Here, what is assumed to be "reality" vanishes.

- *5. Valley of **Unity**,* where the Wayfarer realizes that everything is connected and that the Beloved is beyond everything, including harmony, multiplicity, and eternity.
- *6. Valley of **Wonderment**,* where, entranced by the beauty of the Beloved, the Wayfarer becomes perplexed and, steeped in awe, finds that he or she has never known or understood anything.
- *7. Valley of Poverty and **Annihilation**,* where the self disappears into the universe and the Wayfarer becomes timeless, existing in both the past and the future.

When the birds hear the description of these valleys, they bow their heads in distress; some even die of fright right then and there. But despite their trepidations, they begin the great journey. On the way, many perish of thirst, heat or illness, while others fall prey to wild beasts, panic, and violence. Finally, only thirty birds make it to the abode of Simorgh. In the end, the birds learn that they themselves, united together, are the Simorgh. "Simorgh" in Persian means thirty birds (si means thirty and morgh means bird). They eventually come to realize that the majesty of that Beloved is like the sun that can be seen reflected in a mirror: yet, whoever looks into that mirror will also behold his or her own self.

Attar's masterful use of symbolism is a key driving component of the poem. This adroit handling of symbolisms and allusions can be seen reflected in these lines:

> It was in China, late one moonless night,
>
> The Simorgh first appeared to mortal sight.

Beside the symbolic use of the Simorgh, the allusion to China is also significant. China as used here, is symbolism, as inferred from the Hadith "Seek knowledge even if it be (far off) in China". There are many more examples of such subtle symbols and allusions throughout. Within the larger context of the story of the journey of the birds, Attar masterfully tells the reader many didactic short, sweet stories in captivating poetic style.

Tadhkirat-ul-Awliyā, is a hagiographic collection of Sufis; it is Attar's only known prose work. Written and compiled throughout much of his life and published before his death, the compelling account of the execution of the mystic Mansur al-Hallaj, who had uttered the words "I am the Truth" in a state of ecstatic contemplation, is perhaps the most well known extract from the book.

The **Ilāhī-Nāma** (Persian: الهی‌نامه) is another famous poetic work of Attar, consisting of 6500 verses. In terms of form and content, it has some similarities with Manṭiq-uṭ-Ṭayr. The story is about a king who is confronted with the materialistic and worldly demands of his six sons. The first son asks for the daughter of the king of the fairies, the second for the mastery of magic, the third for the cup of Jamshid, which has the property of displaying the whole world, the fourth for the water of life, the fifth for the ring of Solomon, which has control over fairies and demons, and the sixth for mastering alchemy.

The King tries to show the temporary and senseless desires of his sons by retelling them a large number of spiritual stories. Each of these desires is discussed first literally, and shown to be absurd, and then it is explained how there is an esoteric interpretation of each one.

Mukhtār-Nāma (Persian: مختارنامه), a wide-ranging collection of quatrains, 2088 in number. A coherent group of mystical and religious subjects is outlined. Examples of these subjects are: search for union, sense of uniqueness, distancing from the world, annihilation, amazement, pain, awareness of death. There is an equally rich group of themes typical of lyrical poetry of inspiration adopted by sufi literature; the examples of such themes are: the torment of love, impossible union, beauty of the loved one, stereotypes of the love story, crying in the torment of separation.

The **Diwan** of Attar (Persian: دیوان عطار) consists almost entirely of poems in the Ghazal form, for which no corresponding genre exists in English poetry. He collected his Ruba'iyat ("quatrains") in a separate work called the **Mokhtar-nama**. There are also some **Qasidas** ("Odes"), but they amount to less than one-seventh of the Divan. His Qasidas expound upon mystical and ethical themes and moral precepts. They are sometimes modelled after Sanai. The Ghazals often seem from their outward vocabulary: love and wine poetry, and outcast behavior; the apparently love poems generally imply spiritual experiences of love in the familiar symbolic language of worldly themes of love, drunkenness and insaneness.

Biographical Summary

Attar was born to a Persian family and he practiced the profession of perfumist and personally attended to a very large number of customers. He is mentioned by only two of his contemporaries, `Awfi and Tusi. However, all sources confirm that he was from Nishapur, a major city of Khorasan (now located in the northeast of Iran), and according to `Awfi, he was a poet of the Seljuq period.

Rumi has mentioned: "Attar was the spirit, Sanai his eyes twain, And in time thereafter, Came we in their train". In another poem Rumi mentions:

Attar travelled through all the seven cities of love
While I am only at the bend of the first alley.

Attar's family seems well to do and he received an excellent education in various fields. His works say little about his life. The people he helped in the perfumery used to confide their troubles in Attar and this affected him deeply. Eventually, he abandoned his business and travelled widely – to

Baghdad, Basra, Kufa, Mecca, Medina, Damascus, Khwarizm, Turkistan, and India, meeting with Sufis - and returned promoting Sufi ideas. From childhood onward, encouraged by his father, Attar was interested in the Sufis and their sayings and way of life.

At the age of 78, Attar died a violent death in the massacre which the Mongols inflicted on Nishapur in April 1221. Today, his mausoleum is located in Nishapur. It was built by Ali-Shir Nava'I in the 16th century and later underwent a total renovation during the rule of Reza Shah in 1940.

76. Ahmad Al-Buni

Sharaf al-Din or Shihab al-Din or Muḥyi al-Din Abu al-Abbas Aḥmad ibn Ali ibn Yusuf al-Qurashi al-Sufi, more known as Ahmad al-Buni

(Arabic: أحمد البوني),

(born in Buna, in present-day Annaba, Algeria - died 1225),

was an Algerian mathematician and philosopher and a well-known Sufi and writer on the esoteric value of letters and topics relating to mathematics, sihr (sorcery), and ruhaniyat.

Scientific Contributions

Ahmad Al-Buni wrote the following works among others.

- Shams al-Maʿārif al-Kubrā (The Great Sun of Gnoses), Cairo, 1928.
- Sharḥ Ism Allāh al-aʿẓam fī al-rūḥānī, printed in 1357 AH or in Egypt al-Maṭbaʿa al-Maḥmudiyya al-Tujjariyya bi'l-Azhar.
- Kabs al-iktidā, Oriental Manuscripts in Durham University Library.
- Berhatiah, Ancient Magic Conjuration Of Power.
- Treatise on the Magical Uses of the Ninety-nine Names of God in the Khalili Collection of Islamic Art.

In c. 1200, Ahmad al-Buni showed how to construct magic squares using a simple bordering technique. Al-Buni wrote about Squares and constructed, for example, 4 x 4 Squares using letters from one of the 99 names of Allah.

His works on traditional healing remains a point of reference among Yoruba Muslim healers in Nigeria and some other areas of the Muslim world.

Instead of sihr (black or white magic), this kind of magic was called Ilm al-Hikmah (Knowledge of the Wisdom), Ilm al-isimiyah (Study of the Divine Names) and Ruhaniyat. Most of the so-called mujarrabât ("time-tested methods") books on mgic in the Muslim world are simplified excerpts from the Shams al-ma`ârif. The book remains the seminal work on esotericism.

His work is said to have influenced the Hurufis and the New Lettrist International.

We observe that the work was intended for the specialized audience who could appreciate its import, content and value. However, it became a common availability item, and outside of its domain of the audience, it became grossly misunderstood.

For Muslims, God is a living God who is actively involved in the wellbeing of His creatures. On the other hand, the Muslim Scientists were discovering Laws of Physics, Mathematics, Biology, etc. It then became incumbent upon the people of ruhaniyat to explain how God is active despite the operations of these laws. Works like Shams al-Maʿārif were researched under these requirements and circumstances. Considering their very purpose, these works were only for the eyes of the people advanced in ruhaniyat. They were never to be exposed to the general audience, and when they were, they produced apparent anomalies.

We observe, also, that the work was declared heretical way after the death of Ahmad Al-Buni. During his life, he was regarded with reverence. He was a contemporary of such giants as Ibn-al-Arabi, and no such negative criticism is visible during the lifetime of Ahmad Al-Buni.

Biographical Summary

Ahmad Al-Buni was born in Buna, in present-day Annaba, Algeria. He died in 1225 AD. A contemporary of Ibn Arabi, he is best known for writing one of the most important books of his era; the "Shams al-Ma'arif", a book that is still regarded as the foremost occult text on talismans and divination. It was banned soon after as heretical.

77. Abd al-Salam ibn Mashish al-Alami

'Abd al-Salām ibn Mashīsh al-'Alamī

(Arabic: عبد السلام بن مشيش العلمي),

(1140 – 1227 AD),

was a Moroccan Sufi who lived during the reign of the Almohad Caliphate.

Scientific Contributions

He is the author of a collection of reflections about religious and political life in his time and of a famous eulogy of the prophet Mohammed (taṣliyah) on which a commentary was written by Ahmad ibn Ajiba.

He also wrote a metaphysical paraphrase of a widely known prayer, called al-Salat al-Mashishiyah, in which the believer calls on God to bless the Prophet to thank him for having received Islam through him. In it, Ibn Mashish sees in the prophet Muhammad as a creature of Allah from which all revelation comes and which is the eternal mediator between the ungraspable God and the world.

See Titus Burckhardt, "The Prayer of Ibn Mashish", Studies in Comparative Religion, Winter-Spring, 1978, Pates Manor, Bedfont, Middlesex.

Biographical Summary

Virtually nothing is known about him except that he was born in Jabal Alam in 1140 AD, and he was assassinated in Jabal Alam in 1227 by the anti-Almohad rebel, Ibn Abi Tawajin.

His genealogy was traced through several ancestors - some of them with typically Berber names - all the way to the Prophet of Islam, Muhammad. It is said that he was born to the Banu Arus tribe in the neighborhood of the Jabal al-'Alam, and that at the age of 16 he travelled to the east to study. On his return, in bijaya (Béjaïa), he followed the instructions of the Andalusian mystic Abu Madyan. He came back to stay in his native country, where he withdrew to the mountain to live an edifying life as an ascetic.

He was the spiritual guide of Abu-l-Hassan ash-Shadhili, his only disciple.

78. Ibn Arabi

Muhyī al-Dīn Abū ʿAbd Allāh Muḥammad ibn ʿAlī ibn Muḥammad ibn al-ʿArabī al-Ḥātimī al-Ṭāʾī al-Andalusī al-Mursī al-Dimashqī

(Arabic: محي الدين أبو عبد الله محمد بن علي بن محمد بن العربي الحاتمي الطائي الأندلسي المرسي الدمشقي),

(commonly known as Ibn ʿArabi (Arabic: ابن عربي),

(nicknamed al-Qushayri and Sultan al-ʿArifin),

(1165 – 1240 AD),

was a Sufi, poet, and philosopher, influential within ruhani thought.

Scientific Contributions

Muḥyiddin ibn Arabi is considered a Wali. His cosmological teachings became the dominant worldview in many parts of the Muslim world. He is renowned among practitioners of Sufism by the name al-Shaykh al-Akbar ("the Greatest Shaykh").

Some 800 works are attributed to Ibn Arabi, although not all have been authenticated. Recent research suggests that over 100 of his works have survived in manuscript form, although most printed versions have not yet been critically edited and include errors.

Osman Yahya compiled a definitive bibliography of Ibn Arabi's works; out of the 850 works attributed to him, some 700 are widely considered authentic, while over 400 are still extant. Following are some of Ibn Arabi's works.

- Al-Futūḥāt al-Makkiyya (The Meccan Illuminations), his largest work in 37 volumes originally and published in 4 or 8 volumes in modern times, discussing a wide range of topics from ruhaniyat to Sufi practices and records of his dreams/visions. It totals 560 chapters. In modern editions it amounts to some 15 000 pages.

- Fusus al-Hikam (The Ringstones of Wisdom) (also translated as The Bezels of Wisdom), Composed during the later period of Ibn 'Arabi's life, the work is sometimes considered his most important and can be characterized as a summary of his teachings and ruhani beliefs. It deals with the role played by various prophets in divine revelation.
- The Dīwān, his collection of poetry spanning five volumes, mostly unedited. The printed versions available are based on only one volume of the original work.
- Rūḥ al-quds (The Holy Spirit) a treatise on the soul which includes a summary of his experience from different spiritual masters in the Maghrib. Part of this has been translated as Sufis of Andalusia, reminiscences and ruhani anecdotes about many interesting people whom he met in al-Andalus.
- Mashāhid al-Asrār (Contemplation of the Holy Mysteries), probably his first major work, consisting of fourteen visions and dialogues with God.
- Mishkāt al-Anwār (Divine Sayings), an important collection made by Ibn 'Arabī of 101 hadīth qudsī.
- al-Fanā' fi'l-Mushāhada (The Book of Annihilation in Contemplation), a short treatise on the meaning of mystical annihilation (fana).
- Awrād (Devotional Prayers), a widely read collection of fourteen prayers for each day and night of the week.
- Risālat al-Anwār (Journey to the Lord of Power), a detailed technical manual and roadmap for the "journey without distance".
- Ayyām al-Sha'n (The Book of God's Days), a work on the nature of time and the different kinds of days experienced by gnostics.

- 'Unqā' Mughrib (The Fabulous Gryphon of the Islamic West), a book on the meaning of being a Wali and its culmination in Jesus and the Mahdī.
- al-Ittihād al-Kawnī (The Universal Tree and the Four Birds), a poetic book on the Complete Human and the four principles of existence.
- 'al-Dawr al-A'lā (Prayer for Spiritual Elevation and Protection), a short prayer which is still widely used in the Muslim world.
- Tarjumān al-Ashwāq (The Interpreter of Desires), a collection of nasībs which, in response to critics, Ibn Arabi republished with a commentary explaining the meaning of the poetic symbols.
- At-Tadbidrat al-ilahiyyah fi islah al-mamlakat al-insaniyyah (Divine Governance of the Human Kingdom).
- Hilyat al-abdāl (The Four Pillars of Spiritual Transformation) a short work on the essentials of the ruhaniyat.

Although Ibn Arabi stated on more than one occasion that he did not blindly follow any one of the schools of Islamic mujtahids, he was responsible for copying and preserving books of the Zahirite or literalist school, to which there is fierce debate whether or not Ibn Arabi followed that school.

On an extant manuscript of Ibn Ḥazm, as transmitted by Ibn ʿArabī, Ibn ʿArabī gives an introduction to the work where he describes a vision he had:

"I saw myself in the village of Sharaf near Siville; there I saw a plain on which rose an elevation. On this elevation the Prophet stood, and a man whom I did not know, approached him; they embraced each other so violently that they seemed to interpenetrate and become one person. Great brightness concealed them from the eyes of the people. 'I would like to know,' I thought, 'who is this strange man.' Then I heard someone say: 'This is the Ahl al-Hadith ʿAlī Ibn Ḥazm.' I had never heard Ibn Ḥazm's name before. One of my shaykhs, whom I questioned, informed me that this man is an authority in the field of science of Hadeeth."

Ibn Arabi did delve into specific details at times, and was known for his view that religiously binding consensus could only serve as a source of sacred law if it was the consensus of the first generation of Muslims who had witnessed revelation directly.

Ibn Arabi also expounded on Sufi Allegories of the Sharia, building upon previous work by Al-Ghazali and al-Hakim al-Tirmidhi.

The doctrine of perfect man (**Al-Insān al-Kāmil**) is popularly considered an honorific title attributed to Muhammad having its origins in Islamic ruhaniyet, although the concept's origin is controversial and disputed. Ibn Arabi may have first coined this term in referring to Adam as found in his work *Fusus al-hikam*, explained as an individual who binds himself with the Divine and creation.

Taking an idea already common within Sufi culture, Ibn Arabi applied deep analysis and reflection on the concept of a perfect human and one's pursuit in fulfilling this goal. In developing his explanation of the perfect being, Ibn Arabi first discusses the issue of oneness through the metaphor of the mirror.

In this philosophical metaphor, Ibn Arabi compares an object being reflected in countless mirrors to the relationship between God and his creatures. God's essence is seen in the existent human being, as God is the object and human beings the mirrors. Meaning two things; such that since humans are mere reflections of God there can be no distinction or separation between the two and, without God the creatures would be non-existent.

When an individual understands that there is no separation between human and God they begin on the path of ultimate oneness.

The one who decides to walk in this oneness pursues the true reality and responds to God's longing to be known. The *search within* for this reality of oneness causes one to be reunited with God, as well as, improve self-consciousness.

The perfect human, through this developed self-consciousness and self-realization, prompts divine self-manifestation.

This causes the perfect human to be of both divine and earthly origin. Ibn Arabi metaphorically calls him an Isthmus. Being an Isthmus between heaven and Earth, the perfect human fulfills God's desire to be known. God's presence can be realized through him by others. Ibn Arabi expressed that through self manifestation one acquires divine knowledge, which he called the primordial spirit of Muhammad and all its perfection. Ibn Arabi details that the perfect human is of the cosmos to the divine and conveys the divine spirit to the cosmos.

Ibn Arabi further explained the perfect man concept using at least twenty-two different descriptions and various aspects when considering the Logos. He contemplated the Logos, or "Universal Man", as a mediation between the individual human and the divine essence.

Ibn Arabi believed Muhammad to be the primary perfect man who exemplifies the morality of God. Ibn Arabi regarded the first entity brought into existence was the reality or essence of Muhammad (al-ḥaqīqa al-Muhammadiyya), master of all creatures, and a primary role-model for human beings to emulate. Ibn Arabi believed that God's attributes and names are manifested in this world, with the most complete and perfect display of these divine attributes and names seen in Muhammad. Ibn Arabi believed that one may see God in the mirror of Muhammad. He maintained that Muhammad was the best proof of God and, by knowing Muhammad, one knows God.

Ibn Arabi also described Adam, Noah, Abraham, Moses, Jesus, and all other prophets and various Anbiya' Allah (Muslim messengers) as perfect men, but never tires of attributing lordship, inspirational source, and highest rank to Muhammad.

Ibn Arabi compares his own status as a perfect man as being but a single dimension to the comprehensive nature of Muhammad. Ibn 'Arabi makes extraordinary assertions regarding his own spiritual rank, but qualifying this rather audacious correlation by asserting his "inherited" perfection is only a single dimension of the comprehensive perfection of Muhammad.

His best-known book, entitled 'al-Futuhat al-Makkiyya' (The Meccan Victories or Illuminations) which begins with a statement of doctrine (belief) about which al-Safadi (d. 1363) said:

> "I saw (read) that (al-Futuhat al-Makkiyya) from beginning to end. It consists of the doctrine of Abu al-Hasan al-Ash'ari without any difference (deviation) whatsoever."

Ibn Arabi began writing Futūḥāt al-Makkiyya after he arrived in Mecca in 1202. After almost thirty years, the first draft of Futūḥāt was completed in December 1231, and Ibn Arabi bequeathed it to his son. Two years before his death, Ibn ʻArabī embarked on a second draft of the Futūḥāt in 1238, which included a number of additions and deletions as compared with the previous draft, that contains 560 chapters. The second draft, which is the most widely circulated and used, was bequeathed to his disciple, Sadr al-Din al-Qunawi.

There are many attempts to translate this book from Arabic into other languages, but there is no complete translation of Futūḥāt al-Makkiyya to this day.

There have been many commentaries on Ibn 'Arabī's Fuṣūṣ al-Ḥikam: Osman Yahya named more than 100. The first one was Kitab al-Fukūk written by Ṣadr al-Dīn al-Qunawī who had studied the book with Ibn 'Arabī; the second by Qunawī's student, Mu'ayyad al-Dīn al-Jandi, which was the first line-by-line commentary; the third by Jandī's student, Dawūd al-Qaysarī, which became very influential in the Persian-speaking world.

A recent English translation of Ibn 'Arabī's own summary of the Fuṣūṣ, Naqsh al-Fuṣūṣ (The Imprint or Pattern of the Fusus), as well a commentary on this work, by 'Abd al-Raḥmān Jāmī. Also Naqd al-Nuṣūṣ fī Sharḥ Naqsh al-Fuṣūṣ (1459), by William Chittick was published in Volume 1 of the Journal of the Muhyiddin Ibn 'Arabi Society (1982).

The Fuṣūṣ was first critically edited in Arabic by 'Afīfī (1946) that became the standard in scholarly works. Later in 2015, Ibn al-Arabi Foundation in Pakistan published the Urdu translation, including the new critical of Arabic edition.

An English translation was done in partial form by Angela Culme-Seymour, from the French translation of Titus Burckhardt as Wisdom of the Prophets (1975). A full translation was done by Ralph Austin as Bezels of Wisdom (1980). There is also a complete French translation by Charles-Andre Gilis, entitled Le livre des chatons des sagesses (1997). The only major commentary to have been translated into English so far is by Ismail Hakki Bursevi. It is a translation and commentary in to English of Fusus al-hikam by Muhyiddin Ibn 'Arabi; the English translation is from "Ottoman Turkish by Bulent Rauf in 4 volumes (1985–1991)".

In Urdu, the most widespread and authentic translation was made by Muhammad Abdul Qadeer Siddiqi Qadri, the former Dean and Professor of Theology of the Osmania University, Hyderabad. It is due to this reason that his translation is in the curriculum of Punjab University. Maulvi Abdul Qadeer Siddiqui has made an interpretive translation and explained the terms and grammar while clarifying the Ibn Arabi's opinions. A new edition of the translation was published in 2014 with brief annotations throughout the book for the benefit of contemporary Urdu reader.

Biographical Summary

'Abū 'Abdullāh Muḥammad ibn 'Alī ibn Muḥammad ibn `Arabī al-Ḥātimī aṭ-Ṭā'ī (أبو عبد الله محمد ابن علي ابن محمد ابن العربي الحاتمي الطائي) was a Sufi, poet, and Arab philosopher from the Tayy tribe. He was born in Murcia, Al-Andalus, on the 17th of Ramaḍān 560 AH (28 July 1165 AD).

Ibn Arabi was Sunni, although his writings on the Twelve Imams were also popularly received among Shia. It is debated whether or not he ascribed to the Zahiri madhab which was later merged with the Hanbali school.

After his death, Ibn Arabi's teachings quickly spread throughout the Islamic world. His writings were not limited to the Muslim elites, but made their way into other ranks of society through the widespread reach of the Sufi orders.

Arabi's work also popularly spread through works in Persian, Turkish, and Urdu. Many popular poets were trained in the Sufi orders and were inspired by Arabi's concepts.

Other scholars in his time like al-Munawi, Ibn 'Imad al-Hanbali and al-Fayruzabadi all praised Ibn Arabi as "A righteous friend of Allah and faithful scholar of knowledge", "the absolute mujtahid without doubt" and "the imam of the people of shari'a both in knowledge and in legacy, the educator of the people of the way in practice and in knowledge, and the shaykh of the shaykhs of the people of truth through spiritual experience (dhawq) and understanding".

Ibn Arabi's paternal ancestry was from the South Arabian tribe of Tayy, and his maternal ancestry was North African Berber. Al-Arabi writes of a deceased maternal uncle, Yahya ibn Yughan al-Sanhaji, a prince of Tlemcen, who abandoned wealth for an ascetic life after encountering a Sufi mystic.

His father, 'Ali ibn Muḥammad, served in the Army of Abū ʿAbd Allāh Muḥammad ibn Saʿd ibn Mardanīsh, the ruler of Murcia. When Ibn Mardanīš died in 1172 AD, his father shifted allegiance to the Almohad Sultan, Abū Ya'qūb Yūsuf I, and returned to government service. His family then relocated from Murcia to Seville. Ibn Arabi grew up at the ruling court and received military training.

As a young man Ibn Arabi became secretary to the governor of Seville. He married Maryam, a woman from an influential family.

Ibn Arabi writes that as a child he preferred playing with his friends to spending time on religious education. He had his first vision of God in his teens and later wrote of the experience as "the differentiation of the universal reality comprised by that look". Later he had several more visions of Jesus and called him his "first guide to the path of God. His father, on noticing a change in him, had mentioned this to philosopher and judge, Ibn Rushd (Averroes), who asked to meet Ibn Arabi. Ibn Arabi said that from this first meeting, he had learned to perceive a distinction between formal knowledge of rational thought and the unveiling insights into the nature of things. He then adopted Sufism and dedicated his life to the spiritual path.

When he later moved to Fez, in Morocco, Mohammed ibn Qasim al-Tamimi became his spiritual mentor.

Ibn Arabi left Andalusia for the first time at age 36 and arrived at Tunis in 1193. After a year in Tunisia, he returned to Andalusia in 1194. His father died soon after Ibn Arabi arrived at Seville. When his mother died some months later, he left Andalusia for the second time and travelled with his two sisters to Fez, Morocco in 1195. He returned to Córdoba, Andalusia, in 1198, and left Andalusia crossing from Gibraltar for the last time in 1200. In 1200 he took leave from one of his teachers, Shaykh Abu Ya'qub Yusuf ibn Yakhlaf al-Kumi, then living in the town of Salé.

He received a vision instructing him to journey east. After visiting some places in the Maghreb, he left Tunisia in 1201 and arrived for the Hajj in 1202. He lived in Mecca for three years, and there began writing his work Al-Futūḥāt al-Makkiyya (الفتوحات المكية) – 'The Meccan Illuminations'.

After spending time in Mecca, he traveled throughout Syria, Palestine, Iraq and Anatolia.

In 1204, Ibn Arabi met Shaykh Majduddīn Isḥāq ibn Yūsuf (شيخ مجد الدين إسحاق بن يوسف), a native of Malatya and a man of great standing at the Seljuk court. This time Ibn Arabi was travelling north; first they visited Medina and in 1205 they entered Baghdad. This visit offered him a chance to meet the direct disciples of Shaykh 'Abd al-Qādir Jīlānī. Ibn Arabi stayed there only for 12 days because he wanted to visit Mosul to see his friend 'Alī ibn 'Abdallāh ibn Jāmi', a disciple of the mystic Qaḍīb al-Bān (1079-1177 AD; قضيب البان). There he spent the month of Ramaḍan and composed Tanazzulāt al-Mawṣiliyya (تنزلات الموصلية), Kitāb al-Jalāl wa'l-Jamāl (كتاب الجلال و الجمال), "The Book of Majesty and Beauty", and Kunh mā lā Budda lil-MurīdMinhu.

In the year 1206 Ibn Arabi visited Jerusalem, Mecca and Egypt. It was his first time that he passed through Syria, visiting Aleppo and Damascus.

Later in 1207 he returned to Mecca where he continued to study and write, spending his time with his friend Abū Shujā bin Rustem and his family, including Niẓām.

The next four to five years of Ibn Arabi's life were spent in these lands and he also kept travelling and holding the reading sessions of his works in his own presence.

On 22 Rabīʿ al-Thānī 638 AH (8 November 1240) at the age of seventy-five, Ibn Arabi died in Damascus.

79. Bahauddin Zakariya

Baha-ud-din Zakariya

(also known as Baha-ul-Haq),

(Persian: بهاؤالدين زكريا)

(c.1170 – 1262),

was a sufi and poet who established the Suhrawardia tariqa in South Asia, becoming one of the influential teachers of tasawwuf of his era.

Scientific Contribution

One of Zakariya's main works is (اوراد شيخ الشيوخ : الاوراد : اوراد سهروردي), Awrad-e-Shaikhush Shuyukh: or simply Al-Awrad : or Awrad-e-Suhrawardi.

Zakariya became a vocal critic of Multan's ruler at the time, Nasir-ud-Din Qabacha, and sided with Iltutmish, the Mamluk Sultan of Delhi when he overthrew Qabacha in 1228. Zakariya's support was crucial for Iltutmish's victory, and so he was awarded the title of Shaikh-ul-Islam by Iltutmish to oversee the state's spiritual matters, in gratitude for his support. Zakariya was also granted official state patronage by the Sultan. This happening is what Suharwardi Tariqa prefers, and does not separate tasawwuf and government.

Zakariya befriended Lal Shahbaz Qalandar - a widely revered Sufi saint from Sindh, and founder of the Qalandariyya order of wandering dervishes. As Shaikh-ul-Islam, Zakariya was able to assuage orthodox Muslims, who were offended by the Lal Shahbaz Qalandar's teachings. Zakariya, Shahbaz Qalandar, Baba Farid and Syed Jalaluddin Bukhari, together became the legendary Haq Char Yaar, or "Four friends" group, which is highly revered among South Asian Muslims.

Zakariya's Tariqa, or Sufi philosophical orientation, was to the renowned Persian Sufi master Shahab al-Din Abu Hafs Umar Suhrawardi of Baghdad. The Suhrawardi tariqa rejected a life of poverty, as espoused by the Chisti order that was more prevalent in the Lahore region. Instead, the Suhrawardis believed in ordinary food and clothing, and rejected the Chisti assertion that tasawwuf lay upon a foundation of poverty. The Suhrawardis also rejected the early Chisti practice of dissociation from the political State.

Zakariya's preachings emphasized the need to conform to usual Islamic practices like salat, saum and zakat, but also advocated a philosophy of scholarship (acquiring ilm) combined with tasawwuf. His emphasis on teaching all humans, regardless of class or ethnicity, set him apart from his contemporary Hindu mystics.

He did not reject the traditional sufi music that was heavily emphasized in Chisti worship, but only partook in it on occasion. He rejected the Chisti tradition of bowing in reverence to religious leaders - a practice that may have been borrowed from Hinduism.

Zakariya's teachings spread widely throughout southern Punjab and Sindh, and drew large numbers of converts from Hinduism. His successors continued to exert strong influence over southern Punjab for the next several centuries, while his order spread further east into regions of northern India, especially in Gujarat and Bengal.

Biographical Summary

Baha-ud-din Zakariya was born in 1161 or 1182. His family was of Hashimite lineage, and thus traced their descent back to Asad ibn Hashim, one of the ancestors of the Islamic prophet Muhammad. Baha al-Din's family was originally from the Khwarazm region in Central Asia, but had settled in Kut Karur in the Punjab region, near the city of Multan. His father was Wajih al-Din Muhammad, while his mother was the daughter of Husam al-Din Tirmidhi.

For fifteen years, Zakariya travelled to different cities in southern Punjab, where the sufi order was able to attract large numbers of converts from Hinduism. Zakariya finally settled in Multan in 1222. Under his influence, Multan became known as "Baghdad of the East," and is referred by Zakariya in his Persian poetry:

Multan ma ba jannat a'la barabara
Ahista pa ba-nah ke malik sajda mi kunand.

(Multan of ours is equal to high Paradise; Tread slowly, the angels are in prostration here.)

Baha-ud-Din Zakariya died in 1268 and his mausoleum (Darbar) is located in Multan. The mausoleum is a square of 51 ft 9 in (15.77 m), measured internally. Above this is an octagon, about half the height of the square, above which is a spherical dome. The mausoleum was almost completely ruined during the Siege of Multan in 1848 by the British, but was soon afterward restored by local Muslims. Many pilgrims visit his shrine at the time of his urs from different parts of Pakistan and beyond.

Bahauddin Zakariya University located in Multan is named after him which is a large institution in southern Punjab. Bahauddin Zakaria Express train is named after him, which runs between Karachi and Multan.

80. Ibn Ata Allah al-Iskandari

Tāj al-Dīn Abū'l-Faḍl Aḥmad ibn Muḥammad ibn ʿAbd al-Karīm ibn Abd al-Rahman ibn Abdullah ibn Ahmad ibn Isa ibn Hussein ibn ʿAṭā Allāh al-Judhami al-Iskandarī al-Shādhilī

(died 1309 AD),

was an Egyptian Malikite jurist, muhaddith and the third murshid (spiritual "guide" or "master") of the Shadhili Sufi order.

Scientific Contributions

He was responsible for systematizing Shādhilī doctrines and recording the biographies of the order's founder, Abu-l-Hassan ash-Shadhili, and his successor, Abu al-Abbas al-Mursi. He is credited with having authored the first systematic treatise on dhikr, The Key to Salvation (Miftāḥ al-Falāḥ), but is mostly known for his compilation of aphorisms, the Ḥikam al-ʿAtā'iyya.

Ibn ʿAṭā Allāh was one of those who confronted the controversial theologian Ibn Taymiyya, who was jailed several times for his views on religious issues and for his perceived excesses in attacking the Sufis. His confrontations with Ibn Taymiyya saw Ibn ʿAṭā Allāh leading a procession of some 200 Sufis against Ibn Taymiyya as well as confronting him on issues.

Ibn 'Ata' Allah's works include:

- Kitab al-Hikam (The Book of Wisdom)
- Kitab al-Lata'if fi manaqib Abi l-'Abbas al-Mursi wa Shaykhihi Abi l Hasan (The Subtle Blessings in the Saintly Lives of Abu l-'Abbas al-Mursi and His Master Abu l-Hassan)
- Miftah al-falah wa misbah al-anwah (The Key of Success and the Lamps of Spirits).

- Kitab al-Tanwir fi isqat al-Tadbir (The Illumination on Abandoning Self-Direction)
- Al-Qasd al-mujarrad fi ma'rifat al-Ism al-Mufrad (The Pure Goal Concerning Knowledge of the Unique Name)
- Taj al-arus al-hawi li-tahdhib an-nufus (The Bride's Crown Containing the Discipline of Souls)
- Unwan at-tawfiq fi adab at-tariq (The Sign of Success Concerning the Discipline of the Path)

The wide circulation of Ibn ʿAṭā Allāh's written works led to the spread of the Shādhilī order in North Africa, where the order's founder had been rejected in earlier attempts. The Wafai Sufi order was also derived from his works.

Commentaries on the Ḥikam have been made by some of the most famous masters of the Shadhili order such as Ibn Abbad al-Rundi, Ahmad Zarruq and Ahmad ibn Ajiba as well as non-Shadhilis like the Syrian Islamic law Professor Sa'id Ramadan al-Bouti. A modern English translation of Ḥikam by Muhammed Nafih Wafy was published under the title "The Book of Aphorism" by Islamic Book Trust in Malaysia in 2010.

Biographical Summary

He was born in Alexandria and taught at both the al-Azhar Mosque and the Mansuriyyah madrasa in Cairo. He died in 1309 in Cairo.

81. Ismail Qureshi al Hashmi

Makhdoom Shaikh Imaduddin Ismail Qureshi (Quraishi) Asadi al Hashmi (1260 – 1349),

was a Suharwardi Sufi, and one of the pioneers of Islamic preachers in Allahabad district, India.

Scientific Contributions

Shah Ismail never allowed his name and works to be highlighted in books and discourses, so is the case that except a very few people, not many know about him even in Allahabad, India.

He erected a mosque over a small mound exactly on the banks of river Ganges. This became the center for his teaching and education activities, over the next half century. Seekers from all areas including Kara, Manikpur, Zafarabad, Jaunpur, Jhunsi, Awadh, thronged to be his student. Though a lot of students were his disciples, but he authorized only three students with Khailafat: One Shaikh Abdul Rahim, Second Shaikh Ali, and third Sayyid Muhammad.

Sayyid Muhammad has an alias of Shah Karak; he eventually rose to become Shah Ismail's most famous disciple; and became popularly known as Shah Karak Majzoob (Abdal) of Kara Manikpur, Allahabad. Shah Karak is buried at Kara and is the most famous saint of Allahabad. Khwaja Karak is alternatively known as Khwaja Gurg of Kara.

Khwaja Karak is more known than his master and mentor, Shak Ismail. Research is being undertaken, for example the following citations.

- Allahabd - A Gazetteer by C H Nevill, 1911, Govt Press, United Provinces.
- Akhbar al Akhyar by Shaikh Abdul Haq Muhaddith Dehlavi, 16th Century. Urdu Edition 1990.

- Manba al Ansaab by Sayyid Muin al Haqq Jhoonsvi, Manuscript, British Library - India Office Collections, London.
- Asral al Makhdumein by Shaikh Karim Yar, 1893, Fatehpur.
- Mirat al Asrar by Shaikh Abdul Rahman Chishti, 1648, Printed Maktaba Jam e Noor, Delhi 1997.
- Tarikh Aina e Awadh by Sayyid Shah Abul Hasan Qutubi Manikpuri, 1887, Nizami Press, Cawnpore.
- Tazkirat al Makhdumeen by Khalid Bin Umar, 2004, Manuscript. This is about Pargana Chail and the Sufi saints resting in Pargana Chail.

> Many Sufis used music as a tool for teaching and education. But Shah Ismail strictly adhered to the basic tenets of Islam. He did not allow anything outside Shariah. He did not approve of singing and dancing in the name of Sam'a.

He was laid to rest near his mosque.

> No dome and tomb kind of structure was erected, per his own wish. No Urs rituals take place at his grave in accordance with the instructions of Shariah.

He is the only Saint in whole of North India where no Urs rituals are performed.

Biographical Summary

Shah Ismail is the grandson of Shaikh Bahauddin Zakaria Multani and son of Shaikh Sadruddin Arif Multani. He is commonly known as Makhdoom Shah of Bamrauli.

Shaikh Ismail was born in 1260 AD in Multan, and was brought up under the guidance of his father Shaikh Sadruddin, and grandfather Shaikh Bahauddin Zakaria.

He was initiated into Suharwardi order at the hands of his grandfather Shaikh Bahauddin Zakaria, and trained by his father, Shaikh Sadruddin, and his elder brother, Shaikh Ruknuddin Abul Fateh, alias Shah Rukn e Alam of Multan.

Shah Rukn e Alam kept his younger brother, Shaikh Ismail, always with him and even in his visits to Delhi at the call of Alauddin Khalji. When Shah Rukn e Alam visited Shaikh Nizamuddin Auliya, his younger brother, Shaikh Ismail, was with him.

Shaikh Ismail used to live mainly at Multan looking after the educative activities of his Khanqaah; however, before the death of his elder brother, Shah Rukn e Alam, a divine intuition propelled him to migrate and shift to Allahabad which was called Prayag at that time.

He travelled first to Delhi where the emperor Alauddin Khalji and his successor welcomed him. From there he travelled to Allahabad via kara and settled in a village named Bamhrauli or Bamrauli (Chail, Allahabad). Thus, he is considered to be one of the earliest Shaikhs who settled in the environs of the present city of Allahabad.

Shaikh Ismail strictly adhered to the basic tenets of Islam. He did not like anything outside Shariah. He did not approve of singing and dancing in the name of Sam'a. He exhorted all to live a life of piety respecting their parents and fulfilling the duty by carrying out the responsibilities as the slave and vicegerent of Allah.

He erected a mosque over a small mound on the banks of river Ganges. After spending more than 50 years in that village and commanding respect and honor from all, he died silently in 1349 AD. He was laid to rest near his mosque. No big dome and tomb kind of structure was erected, per his wish. No Urs rituals take place at his grave in accordance with the instructions of Shariah. He is the only Saint in whole of North India where no Urs rituals are performed. This is striking evidence of saint's adherence and strict following of the Shariah.

The writer of Mirat al Asrar paid a special visit to this mosque and tomb in the reign of Shahjahan and commented "that it is a great place of Barkat".

82. Ibn Qayyim al-Jawziyya

Shams al-Dīn Abū ʿAbd Allāh Muḥammad ibn Abī Bakr ibn Ayyūb al-Zurʿī al-Dimashqī al-Ḥanbalī, commonly known as Ibn Qayyim al-Jawziyya ("The son of the principal of the school of Jawziyyah")

(Arabic: محمد بن أبي بكر بن أيوب بن سعد بن حريز بن مكي زين الدين الزُّرعي),

or for short Ibn al-Qayyim ("Son of the principal") (ابن القيّم),

(29 January 1292 – 15 September 1350 CE),

was an important faqih, mutakallim, and ruhani writer.

Scientific Contributions

Ibn Qayyim Al-Jawziyya wrote a lengthy ruhani commentary on a treatise written by the Hanbali Sufi Khwaja Abdullah Ansari entitled Madarij al-Salikin.

Ibn Qayyim al-Jawziyyah's contributions to the Islamic library are extensive, and they particularly deal with the Qur'anic commentaries, and understanding and analysis of the prophetic traditions (Fiqh-us Sunnah) (فقه). He "wrote about a hundred books", including the following:

- Zad al-Ma'ad (Provision of the hereafter)
- Al-Waabil Sayyib minal kalim tayyib – a commentary on hadith about Prophet Yahya ibn Zakariyya.
- I'laam ul Muwaqqi'een 'an Rabb il 'Aalameen (Information for Those who Write on Behalf of the Lord of the Worlds)
- Tahthib Sunan Abi Da'ud
- Madaarij Saalikeen which is an extensive commentary on the book by Shaikh Abu Ismail al-Ansari al-Harawi al-Sufi, Manazil-u Sa'ireen (Stations of the Seekers);
- Tafsir Mu'awwadhatain (Tafsir of Surah Falaq and Nas);
- Badāʾiʿ al-Fawāʾid (بدائع الفوائد): Amazing Points of Benefit

- Ad-Dā'i wa Dawā also known as Al Jawābul kāfi liman sa'ala 'an Dawā'i Shaafi
- Haadi Arwah ila biladil Afrah
- Uddat as-Sabirin wa Dhakhiratu ash-Shakirin (عدة الصابرين وذخيرة الشاكرين)
- Ighathatu lahfaan min masaa'id ash-shaytan (إغاثة اللهفان من مصائد الشيطان): Aid for the Yearning One in Resisting the Shayṭān
- Rawdhatul Muhibbīn
- Ahkām ahl al-dhimma"
- Tuhfatul Mawdud bi Ahkam al-Mawlud: A Gift to the Loved One Regarding the Rulings of the Newborn
- Miftah Dar As-Sa'adah
- Jala al-afham fi fadhl salati ala khayral anam
- Al-Manar al-Munif
- Al-Tibb al-Nabawi – a book on Prophetic medicine, available in English as "The Prophetic Medicine", printed by Dar al-Fikr in Beirut (Lebanon), or as "Healing with the Medicine of the Prophet (sal allahu `alayhi wa salim)", printed by Darussalam Publications.
- Al-Furusiyya
- Shifa al-Alil fi masa'il al qada'i wal qadri wal hikmati wa at-ta'leel (Remedy for Those who Question on Matters Concerning Divine Decree, Predestination, Wisdom and Causality)
- Mukhtasar al-Sawa'iq
- Hadi al-Arwah ila Bilad al-Arfah (Spurring Souls on to the Realms of Joy
- A treatise on Arab archery is by Ibn Qayyim Al-Jawziyya, Muḥammad ibn Abī Bakr (1292AD-1350AD) and comes from the 14th century.

Like his teacher Ibn Taymiyya, Ibn Qayyim, supported broad powers for the state and prosecution. He argued, for example, "that it was often right to punish someone of lowly status" who alleged improper behavior by someone "more respectable."

Ibn Qayyim "formulated evidential theories" that made judges "less reliant than ever before on the oral testimony." One example was the establishment of a child's paternity by experts scrutinizing the faces of "a child and its alleged father for similarities". Another was in determining impotence. If a woman sought a divorce on the grounds of her husband's impotence and her husband contested the claim, a judge might obtain a sample of the husband's ejaculate. According to Ibn Qayyim "only genuine semen left a white residue when boiled".

In interrogating the accused, Ibn Qayyim believed that testimony could be beaten out of suspects if they were "disreputable". This was in contrast to the majority of Islamic faqihs who had always acknowledged "that alleged sinners were entitled to remain silent if accused." Attorney and author Sadakat Kadri states that, "as a matter of straightforward history, torture had originally been forbidden by Islamic faqihs. "Ibn Qayyim however, believed that "the Prophet Muhammad, the Rightly Guided Caliphs, and other Companions" would have supported his position.

Ibn Qayyim al-Jawziyyah opposed alchemy and divination of all varieties, but was particularly opposed to astrology, whose practitioners dared to "think they could know secrets locked within the mystery of God's supreme and all-embracing wisdom." In fact, those who believed that human personalities and events were influenced by heavenly bodies, were "the most ignorant of people, the most in error and the furthest from humanity ... the most ignorant of people concerning his soul and its creator".

In his Miftah Dar al-Sa'adah, in addition to denouncing the astrologers as worse than infidels, he uses empirical arguments to refute the practice of alchemy and astrology along with the theories associated with them, such as divination and the transmutation of metals, for example arguing that:

> And if you astrologers answer that it is precisely because of this distance and smallness that their influences are negligible, then why is it that you claim a great influence for the smallest heavenly body, Mercury? Why is it that you have given an influence to al-Ra's and al-Dhanab, which are two imaginary points [ascending and descending nodes]?"

Although Ibn al-Qayyim is sometimes characterized today as an unabashed enemy of Islamic mysticism, it is historically known that he actually had a "great interest in Sufism," which arose out of his vast exposure to the practice given Sufism's integral role in orthodox Islamic life at his time. Some of his major works, such as Madārij, Ṭarīq al-hijratayn (Path of the Two Migrations) and Miftāḥ dār al-saʿāda (Key to the Joyous Dwelling), "are devoted almost entirely to Sufi themes," yet allusions to such "themes are found in nearly all his writings," including in such influential works of spiritual devotion such as al-Wābil al-Ṣayyib, an important treatise detailing the importance of the practice of dhikr, and his revered magnum opus, Madārij al-sālikīn (The Wayfarers' Stages), which is an extended commentary on a work written by the eleventh-century Hanbalite saint and mystic Abdullah Ansari, whom Ibn al-Qayyim referred to reverentially as "Shaykh al-Islām."

In all such writings, it is evident Ibn al-Qayyim wrote to address "those interested in Sufism in particular and ... 'the matters of the heart' ... in general," and proof of this lies in the fact that he states, in the introduction to his short book Patience and Gratitude, "This is a book to benefit kings and princes, the wealthy and the indigent, Sufis and religious scholars; (a book) to inspire the sedentary to set out, accompany the wayfarer on the Way (al-sā'ir fī l-ṭariq) and inform the one journeying towards the Goal." Some scholars have compared Ibn al-Qayyim's role to that of Ghazali two-hundred years prior, in that he tried to "rediscover and restate the orthodox roots of Islam's interior dimension."

It is also true, however, that Ibn al-Qayyim did indeed share some of his teacher Ibn Taymiyyah's more negative sentiments towards what he perceived to be excesses in mystical practice. For example, he felt that the pervasive and powerful influence the works of Ibn Arabi had begun to wield over the entire Sunni world was leading to errors in doctrine. As a result, he rejected Ibn Arabi's concept of wahdat al-wajud or the "oneness of being," and opposed, moreover, some of the more extreme "forms of Sufism that had gained currency particularly in the new seat of Muslim power, Mamluk Egypt and Syria." That said, he never condemned Sufism outright, and his many works bear witness, as it has been noted above, to the immense reverence in which he held the vast majority of Sufi tradition.

In this connection, it is also significant that Ibn al-Qayyim followed Ibn Taymiyyah in "consistently praising" the early spiritual master al-Junayd, one of the most famous saints in the Sufi tradition, as well as "other early spiritual masters of Baghdad who later became known as 'sober' Sufis." As a matter of fact, Ibn al-Qayyim did not condemn the ecstatic Sufis either, regarding their mystical outbursts as signs of spiritual "weakness" rather than heresy.

Ibn al-Qayyim's highly nuanced position on this matter led to his composing apologies for the ecstatic outbursts of several early Sufis, just as many Sufis had done so before him.

Ibn Qayyim was respected by a number of scholars during and after his life. Ibn Kathir stated that Ibn al-Qayyim, was the most affectionate person. He was never envious of anyone, nor did he hurt anyone. He never disgraced anyone, nor did he hate anyone. ... I do not know in this world in our time someone who is more dedicated to acts of devotion.

Ibn Rajab, one of Ibn Qayyim's students, stated that, although, he was by no means infallible, no one could compete with him in the understanding of the texts.

Ibn Qayyim was criticized by a number of scholars, including:

- Taqi al-Din al-Subki (d. 1355) accused him of heresy, and wrote a book against him, entitled: "Al-Sayf al-Saqil fi al-Radd ala Ibn Zafil".
- Ibn Hajar al-Haytami (d. 1566–7) in his al-Fatawa al-Hadithiyya declared Ibn al-Qayyim and his teacher Ibn Taymiyya to be heretics and unbelievers (Mulhideen). He described their position on the Divine attributes as anthropomorphist.

Biographical Summary

Muhammad Ibn Abī Bakr Ibn Ayyub Ibn Sa'd Ibn Harīz Ibn Makkī Zayn al-Dīn al-Zur'ī (Arabic: محمد بن أبي بكر بن أيوب بن سعد بن حريز بن مكي زين الدين الزُّرعي), has kunya of Abu Abdullah (أبو عبد الله). He is usually known as Ibn Qayyim al-Jawziyyah, after his father Abu Bakr Ibn Sa'd al-Zur'ī who was the superintendent (qayyim) of the Jawziyyah Madrasah, the Hanbali law college in Damascus.

Ibn al-Qayyim was born in 1292 AD. He studied mainly from Ibn Taymiyyah; though he also studied under a number of other scholars including his father (Abu Bakr ibn Ayoub), Ibn 'Abd Al Da'im, Shams ad-Dīn adh-Dhahabī, and Safi Al-Din Al-Hindi.

Ibn al-Qayyim studied under Ibn Taymiyyah during 1313-1328, after Ibn Taymiyyah moved back to Damascus from Cairo till he died in 1328. Ibn al-Qayyim was imprisoned with his teacher Ibn Taymiyyah from 1326 until 1328, when Ibn Taymiyyah died and Ibn al-Qayyim was released. As a result of this 16-year union, Ibn al-Qayyim shared many of his teacher's views on various issues.

According to the historian al-Maqrizi, two reasons led to his arrest: the first was a sermon Ibn al-Qayyim had delivered in Jerusalem in which he decried the visitation of holy graves, including the Prophet Muhammad's grave in Medina; the second was his agreement with Ibn Taymiyyah's view on the matter of divorce, which contradicted the view of the majority of scholars in Damascus. The campaign to have Ibn al-Qayyim imprisoned was led by Shafi'I and Maliki scholars, and was also joined by the Hanbali and Hanafi judges.

Whilst in prison Ibn al-Qayyim busied himself with the Qur'an. According to Ibn Rajab, Ibn al-Qayyim made the most of his time of imprisonment: the immediate result of his delving into the Qur'an while in prison was a series of mystical experiences (described as dhawq, direct experience of the divine mysteries, and mawjud, ecstasy occasioned by direct encounter with the Divine Reality).

Ibn al-Qayyim died at the age of 60 years, 5 months, and 5 days, on the 13th night of Rajab, 751 AH (September 15, 1350 CE), and was buried besides his father at the Bab al-Saghīr Cemetery.

He was of Hanbali school of orthodox Sunni jurisprudence, of which he is regarded as one of the most important thinkers. Ibn al-Qayyim was also the foremost disciple and student of Ibn Taymiyyah, with whom he was imprisoned in 1326 for dissenting against established tradition during Ibn Taymiyyah's famous incarceration in the Citadel of Damascus.

Ibn al-Qayyim was of humble origin, Ibn al-Qayyim's father was the principal (qayyim) of the School of Jawziyya, which also served as a court of law for the Hanbali judge of Damascus during the time period. Ibn al-Qayyim went on to producing a rich corpus of "doctrinal and literary" works. As a result, numerous important Muslim scholars of the Mamluk period were among Ibn al-Qayyim's students or, at least, greatly influenced by him, including, amongst others, the Shafi historian Ibn Kathir (d.1373), the Hanbali hadith scholar Ibn Rajab (d. 1397), and the Shafi polymath Ibn Hajar al-Asqalani (d. 1449).

Ibn al-Qayyim died in 1350 AD.

In the present day, Ibn al-Qayyim's name has become a controversial one in certain quarters of the Islamic world due to his popularity amongst many adherents of the Sunni movements of Salafism and Wahhabism, who see in him a classical precursor to their own perspective, because of his criticisms of such widespread practices of the veneration of Auliya and the veneration of their graves and relics.

83. Ibn Qudamah

Ibn Qudāmah al-Maqdisī Muwaffaq al-Dīn Abū Muḥammad ʿAbd Allāh b. Aḥmad b. Muḥammad

(Arabic: ابن قدامة المقدسي موفق الدين ابو محمد عبد الله بن احمد بن محمد),

(often referred to as Ibn Qudamah),

(1147 - 7 July 1223),

was a sufi, zahid, and faqih. He is one of the most notable and influential thinkers of the Hanbali school of Sunni fiqh, with the honorific epithet of Shaykh-ul-Islam. Ibn Qudamah is famous for having said: "The consensus of the Imams of fiqh is an overwhelming proof and their disagreement is a vast mercy."

Scientific Contributions

In aqeeda, Ibn Qudamah was one of the primary proponents of the Athari school of Sunni kalam, which held that overt ilm al kalam speculation was spiritually detrimental, and supported drawing kalam exclusively from the two sources: Quran and the hadith.

Regarding kalam, Ibn Qudamah famously said:

> "We have no need to know the meaning of what God - Exalted is He - intended by His attributes - He is Great and Almighty. No deed is intended by them. No obligation is linked to them except belief in them. Belief in them is possible without knowing their meaning."

According to one scholar, it is evident that Ibn Qudamah "completely opposed discussing matters of ilm al kalam and permitted no more than repeating what was said about God in the revelation." In other words, Ibn Qudamah rejected "any attempt to link God's attributes to the referential world of ordinary human language," that is to say, as a ilm al kalam point of view it purposefully avoided any type of speculation or reflection upon the nature of God.

Ibn Qudamah's attitude towards kalam was challenged by certain later Hanbali thinkers like Ibn Taymiyyah (d. 1328), who engages "in interpretation[s] of the meanings of God's attributes."

Ibn Qudamah seems to have been a formidable opponent of deviants in Islamic practice, as is evidenced by his famous words:

> "There is nothing outside of Paradise but hell-fire; there is nothing outside of the truth but error; there is nothing outside of the Way of the Prophet but heretical innovation."

Ibn Qudamah appears to have been a supporter of seeking the intercession of Muhammad in personal prayer, for he approvingly cites the famous prayer attributed to Ibn Hanbal (d. 855):

> "O God! I am turning to Thee with Thy Prophet, the Prophet of Mercy. O Muhammad! I am turning with you to my Lord for the fulfillment of my need."

Ibn Qudama also relates that which al-'Utbiyy narrated concerning one's visitation to the grave of Muhammad in Medina:

I was sitting by the grave of the Prophet, peace and blessings be upon him, when a bedouin man [a'rābī] entered and said, "Peace be upon you, oh Messenger of God. I have heard God say [in the Qur'an], 'Had they come to you [the Prophet] after having done injustice to themselves [sinned] and asked God for forgiveness and [additionally had] the Messenger asked for forgiveness on their behalf, they would have found God to be oft-turning [in repentance] and merciful.' And I have come to you seeking forgiveness for my sin[s], and seeking your intercession near God."

He [the bedouin man] then said the following poem:

O he who is the greatest of those buried in the grandest land,
[Of] those whose scent has made the valley and hills fragrant,
May my life be sacrificed for the grave that is your abode,
Where chastity, generosity and nobility reside!

Al-'Utbiyy then narrates that he fell asleep and saw the Prophet in a dream and was informed that the bedouin man had indeed been forgiven.

After quoting the above event, Ibn Qudamah explicitly recommends that Muslims should use the above prayer when visiting the Prophet. He thus approves of asking the Prophet for his intercession even after his earthly death.

As is attested to by numerous sources, Ibn Qudamah was a devoted mystic and zahid of the Qadhiriyya order of Sufism, and reserved "a special place in his heart for mystics and mysticism" for the entirety of his life. Having inherited the "spiritual mantle" (khirqa) of Abdul-Qadhir Gilani prior to the renowned spiritual master's death, Ibn Qudamah was formally invested with the authority to initiate his own disciples into the Qadhiriyya tariqa. Ibn Qudamah later passed on the initiatic mantle to his cousin Ibrāhīm b. ʿAbd al-Wāḥid (d. 1217), another important Hanbali faqih, who became one of the primary Qadhiriyya spiritual masters of the succeeding generation. According to some classical Sufi chains, another one of Ibn Qudamah's major disciples was his nephew Ibn Abī ʿUmar Qudāmah (d. 1283), who later bestowed the khirqa upon Ibn Taymiyyah, who, as many recent academic studies have shown, actually appears to have been a devoted follower of the Qadhiriyya Sufi order in his own right, despite his criticisms of several of the most widespread, orthodox Sufi practices of his day and, in particular, of the philosophical influence of the Akbari school of Ibn Arabi.

Due to Ibn Qudamah's public support for the necessity of Sufism in orthodox Islamic practice, he gained a reputation for being one of "the eminent Sufis" of his era.

Ibn Qudamah supported using the relics of Muhammad for the deriving of holy blessings, as is evident from his approved citing, in al-Mughnī 5:468, of the case of Abdullah ibn Umar (d. 693), whom he records as having placed "his hand on the seat of the Prophet's minbar ... [and] then [having proceeded to] wipe his face with it." This view was not novel or even unusual in any sense, as Ibn Qudamah would have found established support for the use of relics in the Quran, hadith, and in Ibn Hanbal's well-documented love for the veneration of Muhammad's relics.

Ibn Qudamah staunchly criticized all who questioned or rejected the existence of saints, the veneration of whom had become an integral part of Sunni piety by the time period, and which he endorsed. As ulama have noted, Hanbali authors of the period were united in their affirmation of sainthood and saintly miracles, and Ibn Qudamah was no exception.

Thus, Ibn Qudamah vehemently criticized what he perceived to be the rationalizing tendencies of Ibn Aqil for his attack against the veneration of saints, saying: "As for the people of the Sunna who follow the traditions and pursue the path of the righteous ancestors, no imperfection taints them, nor does any disgrace occur to them."

Following are some of Ibn Qudamah's works:

- Lum`at ul-I`tiqād (The Illuminating Creed).

- Al-Mughnī (The Persuader) is the most advanced book of Ibn Qudamah.
- Kitāb ut-Tawwābīn.
- Ithbāt Sifat il-`Uluww.
- Dhamm ut-Ta'wīl.
- Al-Burhān Fī Bayan Al-Qur'ān.
- Al-'Umdah ("the support"), a beginner's guide to Ḥanbalī Fiqh. A number of commentaries have been written on this including "Sharh Al-`Umdah" of Ibn Taymiyyah.
- Al-Muqni' Fi Fiqh Al-Imam Ahmad Bin Hanbal Ash-Shaybānī.
- "Kitāb Al-Hādī" or "Umdatul-Hazim fi-l Masail al-Zawa-id 'An Mukhtasar Abi-l Qasim".
- Al-Kaafi
- Rawḍat al-Nāẓir, a book on the fundamentals of uṣūl al-fiqh.
- Al-Waşiyyah (The Advice).
- Ar-Riqqatu wal-Bukāe (sensibility and tears)
- Taḥrīm al-Naḍar fī Kutub al-Kalām (The Censure of Speculative Theology).
- Hikāyat ul-Munādhara Fil-Qur'an (a documentation of a debate he had with the Ash'aris on the subject of the Qur'an).
- Muntakhab min al-'ilal lil-Khallāl (المنتخب من العلل للخلال).

Biographical Summary

Ibn Qudamah was born in Palestine in Jammain, a town near Jerusalem in 1147 to the revered Hanbali preacher and mystic Aḥmad b. Muḥammad b. Qudāmah (d. 1162), "a man known for his zuhd" and in whose honor "a mosque was [later] built in Damascus." Having received the first phase of his education in Damascus, where he studied the Quran and the hadith extensively, Ibn Qudamah made his first trip to Baghdad in 1166, in order to study fiqh and Sufi mysticism under the tutelage of the renowned Hanbali mystic and faqih Abdul-Qadhir Gilani (d. ca. 1167); Gilani would go on to become one of the most widely venerated saints in Sunni Islam. Although Ibn Qudamah's discipleship was cut short by the latter's death, [the] experience [of studying under Abdul-Qadhir Gilani] had its influence on young Ibn Qudamah regarding mystics and mysticism.

Ibn Qudamah's first stay in Baghdad lasted four years, during which time he is also said to have written an important work entitled Taḥrīm al-naẓar fī kutub ahl al-kalām (The Censure of Rationalistic kalam), criticizing what he deemed to be the excessive rationalism of Ibn Aqil (d. 1119).

During this sojourn in Baghdad, Ibn Qudamah studied hadith under numerous teachers, including three female hadith masters, namely Khadīja al-Nahrawāniyya (d. 1175), Nafīsa al-Bazzāza (d. 1168), and Shuhda al-Kātiba (d. ca. 1175). All these various teachers gave Ibn Qudamah the permission to begin teaching the principles of hadith to his own students, including important female disciples such as Zaynab bint al-Wāsiṭī (d. ca. 1240).

Ibn Qudamah fought in Salahuddin's Army during the battle to recapture Jerusalem in 1187. He visited Baghdad again in 1189 and 1196, making his pilgrimage to Mecca the previous year in 1195, before finally settling down in Damascus in 1197.

Ibn Qudamah died on Saturday, the Day of Eid al-Fitr, on July 7, 1223.

Mujtahids

A mujtahid is someone who can use ijtihad, which roughly means exercising qiyas and istihsan in an error-free way, and can pass judgement on the matters of other religious affairs in an error-free way. A mujtahid is knowledgeable about the scriptures, their tafsir and ta'wil, the application of qiyas, and limitations of istihsan, etc. In addition, a mujtahid conforms to the scriptures in an exemplary way, and is also exemplary in his inter personal dealings with people. He or she is also an established qiyasist, istihnasist, factologist, muhaddith, and awwalagist, etc. A mujtahid is one who can exercise ijtihad, which is a process that exercises all the above-mentioned elements with exceptional accuracy, almost error free.

A mujtahid exercises all Islamic knowledge to accurately extend it to situations not initially experienced or anticipated. There is no equivalent word for Mujtahid in English language.

Discourse about Ijtehad: Five Imams are popular: Imam Abu Hanifa, Imam Malik, Imam Shafi, Imam Hanbal, and Imam Jaffer Sadiq. However, this epithet "Imam" is only in a lose sense of popular public approval. Their exercise of qiyas and istihsan and other ulum in Islam was not generally accepted widely in learned circles: and that is why we have five of them, and they were almost contemporaries, living over a short period of time. The general Muslim ulama became so concerned with the situation that they refused to welcome any more Imams in fiqh, insisting that five Imams are enough and all people should follow one or another of them. That gave rise to the practice of Taqlid, which caused deterrence to independent research and scholarship among Muslims.

If the five Imams are not Mujtahid, and the door of Ijtehad is closed by the practice of Taqlid, who then is a Mujtahid among Muslims?

In actual history there have been only two Mujtahids: they were the saviors of Islam at two very critical moments in its history. The first was the time when Islam was beginning as a budding effort. At that time Khadija bint Khuwaylid (خَدِيجَة بِنْت خُوَيْلِد) understood Islam like no one else did, except the Prophet himself. She saved Islam during the toughest years Islam ever faced. She demonstrated extremely in depth appreciation of what Islam was, and stood in its support with endurance, sacrifice, courage and steadfastness. The second time it was immediately after the death of the Prophet. At that time Islam was ready to disperse away in Arabia, because many Arab Chieftains thought their allegiance was to Mohammad and it expired upon his death. Abu Bakr stood firm with his unparalleled understanding of Islam. He not only kept Islam from scattering away, he grew its boundaries. This took an unmatched appreciation of what Islam was, and it took extraordinary courage and steadfastness.

Any Mujtahid who the Ulama regard as such would be at best a tiny fraction of what these two towering personalities offered for Islam. Notwithstanding this situation, we have included Izz al-Din Abdul Aziz bin Abd al-Salam as a Mujtahid, because he appears most comprehensively knowledgeable, experienced, a Sufi inclination and pious.

84. Izz al-Din ibn 'Abd al-Salam

Abdul Aziz bin Abdul Salam bin Abi Al-Qasim bin Hassan bin Muhammad bin Mudhahb.

(Also, Abu Muhammad Izz al-Din Abdul Aziz bin Abd al-Salam bin Abi al-Qasim bin Hassan al-Salami al-Shafi'I),

(Arabic: أبو محمد عز الدين عبد العزيز بن عبد السلام بن أبي القاسم بن حسن السُّلَمي الشافعي),

(also known by his titles, Sultan al-'Ulama, Abu Muhammad al-Sulami),

(577 AH - 660 AH / 1262 CE),

was a famous mujtahid, Ash'ari mutakallim, qadhi and the leading Shafi'i authority of his generation.

Scientific Contributions

Izz al-Din ibn 'Abd al-Salam was described by Al-Dhahabi as someone who attained the rank of ijtihad, with asceticism and piety and the command of virtue and forbidding of what is evil and solidity in religion. He was described by Ibn al-Imad al-Hanbali as the sheikh-al-Islam, the imam of the imams, the lone of his era, the authority of ulama, who excelled in fiqh, origins and the Arabic language, and reached the rank of ijtihad, and received students who traveled to him from all over.

Zaki al-Din al-Mundhiri, the Shafi'i faqih, hadith expert and author stated that, "We used to give legal opinions before shaykh 'Izz al-Din arrived; now that he is among us we no longer do so."

Qarafi describes Ibn 'Abd al-Salam as a "staunch defender of the sunna who had no fear of those in power".

A number of sources report that Ibn 'Abd al-Salam reached the level of ijtihad transcending the Shafi'i madhab altogether.

He produced a number of brilliant works in Shafi'i fiqh, Qur'anic awwalogy, methodological fundamentals in fiqh, formal legal opinion, Sufism, and government. His main and enduring contribution was his masterpiece on Islamic legal principles: Qawa'id al-ahkam fi masalih al-anam.

Some of his works are given below:

- Qur'aan
- Tafsir al-Qur'an al-Azim,
- Mukhtasar al-Nukat wa'l 'Uyun lil Imam al-Mawardi,
- Al-Isharah ila al-Ijaz
- Fawa'id fi Mushkil al-Qur'an
- Amali
- Hadith / Sirah
- Mukhtasar Sahih Muslim
- Bidayat al-Sul fi Tafdhil al-Rasul; available in its translated form as The Beginning of The Quest of the High Esteem of the Messenger
- Targhib Ahl al-Islam fi Sakni al-Sham
- Aqeedah
- Al-Mulhat fi I'tiqad Ahl al-Haqq, or by its other title; al-Radd 'ala al-Mubtadi'ah wa'l Hashawiyah; transmitted by his son 'Abd al-Latif.
- Al-Farq bayn al-Iman wa'l Islam or Ma'na al-Iman wa'l Islam
- Al-Anwa' fi 'ilm al-Tawhid
- Bayan Ahwal al-Nas yawm al-Qiyamah
- Tasawwuf / Raqa'iq
- Shajarat al-'Arif wa'l Ahwal wasalih al-Aqwal wa'l A'mal
- Al-Fitan wa'l Balaya wa'l Mihan

- Mukhtasar Ra'ayah al-Muhasibi or Maqasid al-Ri'ayah li Huququillah
- Usool
- Qawa'id al-Kubra or by its full title; Qawa'id al-Ahkam fi Masalih al-Anam. Its popular commentary is available by Imam al-Qarafi who was one of his students.
- Al-Qawa'id al-Sughra, or al-Fawa'id fi Mukhtasar al-Qawa'id; is an abridgement of the above title.
- Al-Imam fi Bayan Adillat al-Ahkam, or ad-Dala'il al-Muta'aliqah bi'l Mala'ikah wa'l Nabiyin
- Fiqh
- Al-Ghayah fi Ikhtisar al-Nihayah; is an abridgement of Nihayat al-Matlab fi Dirayat al-Madhab of imam al-Haramayn al-Juwayni.
- Al-Jam' bayaan al-Hawi wa'l Nihayah; not known to have finished it.
- Al-Fatawa al-Misriyyah, Al-Fatawa al-Musiliyyah, At-Targhib 'a Salat al-Ragha'ib, or by another title; al-Targhib 'an Salat al-Ragha'ib al-Mawdu'ah wa bayan ma fiha min Mukhalafat al-Sunan al-Mashru'ah, or by another title; Risalat fi Dhamm Salat al-Ragha'ib.
- Risalat fi Radd Jawaz Salat al-Ragha'ib or by the title of Risalat fi Tafnid Radd Ibn al-Salah
- Maqasid al-Sawm
- Manasik al-Hajj
- Maqasid al-Salah
- Ahkam al-Jihad wa Fadha'ilihi.

Biographical Summary

Izz al-Din ibn 'Abd al-Salam was born in Damascus in 1181 AD. His full name was Abdul Aziz bin Abdul Salam bin Abi Al-Qasim bin Hassan bin Muhammad bin Mudhahb. He grew up in Damascus and was educated there by such scholars as Ibn Asakir and Jamal al-Din al-Harastani in Sacred law, Sayf al-Din al-Amidi in usul al-Fiqh and kalam, and Tasawwuf with Suhrawardi and Abul Hasan al-Shadhili. He also studied the sciences of the Sharia and the Arabic language.

He preached at the Umayyad Mosque and taught in the corner of Al-Ghazali. He was famous for his knowledge until he reached out to students from the country, which led to his incarceration.

In Damascus, as sermon giver (khatib) of the Umayyad mosque, he openly defied what he considered to be unsanctioned customs followed by the other sermon givers: he refused to wear black, refused to say his sermons in rhymed prose (saj) and refused to praise the princes. When the ruler As-Salih Ismail made capitulatory concessions to Theobald during the Barons' Crusade, Ibn 'Abd al-Salam condemned him from the pulpit and omitted mentioning him in the post-sermon prayer. He was consequently jailed and upon release emigrated to Egypt.

Having left Damascus, Ibn 'Abd al-Salam settled in Cairo where he was appointed chief judge and Imam of the Friday prayer, gaining such public influence that he could (and did) command the right and forbid the wrong with the force of the law.

Ibn 'Abd al-Salam later resigned from the judiciary and undertook a career as a teacher of Shafi'i law at the Salihiyya, a college founded in the heart of Cairo by al-Malik al-Salih which had then barely been completed and which was, in Egypt, the first establishment providing instruction in the four rites. The biographers indicate that he was the first to teach Qur'anic commentary in Egypt.

Ibn 'Abd al-Salam's exploits eventually earned him the title Sultan al-'Ulema (Sultan of the scholars).

He incited people to fight the Mongols and the Crusaders, and participated in jihad himself.

Izz al-Din ibn 'Abd al-Salam died in Cairo in the year 1262 AD.

Concluding Remarks

We have presented 84 scientists in the Religious Sciences from the part 1 (AD 610 to 1400) of the Islamic Era (AD 610 to 1922). All of them are Muslims, an expression of the fact that only Islam admits a scientific treatment of religion. That way is the solidly open path to appreciate the truth in Quran and to approach closer to its Speaker.

A unique example is the science of Ilm-ar-Rijal which the Mohadditheen have developed. It addresses this important question: someone having written or narrated a historical fact, how to ascertain its validity and accuracy? Remarkably, this question has been addressed only by Muslims. European historians have by and large ignored this scientific requirement for the validity and verification of the written word or a verbal narration. Muslims regard these two, the written word and verbal narration, on equal footing; and they invented sciences for this kind of enquiry.

Another example of the religious science is the science of Awwalogy. In this science the meaning of a text is arrived at so that they scientifically incorporate the circumstances (sociocultural, sociopolitical, and linguistic, etc) in which the text or the narration originated. This science is not used in Christian, Buddhist, Hindus and Jewish exegeses. There are multiple reasons: the most practical one is that the original sociocultural, sociopolitical and linguistic considerations are not known in these cases.

The ignorance in the West regarding the sciences of Ilm-ar-Rijal and Awwalogy has generated a lot of fake information by the Church and the Orientalists with respect to Quran and Islam.

A corollary of Ilm-ar-Rijal and Awwalogy is the development of a very fine-tuned judicial science by the Muslims. While Quran provides a constitution for Muslims, the two mentioned sciences have given rise to a judicial science for the use of precedence. The five Imams are considered to be the pioneers of the judicial science. One aspect in these judicial systems is the use of precedence in Islamic Judicial practices. It makes a strict use of science of Factology to validate a precedence. Only the validated precedences are used, and preference is given to the precedences from the early Islamic Era.

Another religious science that Muslims have developed is that of ruhaniyyaat. It invokes an explicit and direct relationship between Islamic rituals and Islamic ruhaniyyaat. As we have said, ruhaniyyaat is squarely an Islamic science.

Among the Muslims, those who focus on the interdependence of the rituals and ruhaniyyaat are given a name, the Sufis. Sufiism is a manifestation of Islamic rituals when they are strongly coupled with ruhaniyyat. It is not what many in the West have hijacked it into; it is not like spiritual but not religious attitudes; it is not the name of isolation, meditation, yoga and seeking paranormal things. It demands unflinching attention for Islamic ways; it is not those who wear wool; it is not any of the stories people might tell you about Hassan al-Basri and Rabia al-Basri. As we said, Sufiism is the practice of Islamic rituals such that Islamic Ruhaniyyat is achieved through them. This is a fine truth about the science of ruhaniyyat.

There are other truths about the religious sciences. One is revealed by a study of fiqhs, the way the subject was treated by the four Sunni Imams versus how it was treated under Mu'tzila and under Imam Zaid bin Ali. This analysis reveals that the Sunni approach prefers to go on after individual efforts; the Mu'tzila downplays the individuals and puts the principles in the foreground, while setting up a tradition of collaborative teamwork; and the example of Imam Zaid bin Ali demonstrates how to be inspired by the Prophet, and to let go of the sense of ownership, and group identity.

The series generally makes it explicit that there is a natural affinity between Islam and science because Quran exhorts its readers to a scientific outlook in life by urging them to observe the nature and the universe around them to get to know the universe. The series includes but only a few from the Muslim science community in the Islamic Era. There are innumerable others, and many have been lost to oblivion. There is a wealth of "science" buried in that community and it remains to be extracted from the archives. Researchers will no doubt make further discoveries. Subsequent editions of this book would expand the set of scientists included, as well as additional details about those already covered.

We have not explicitly stated the inclusion criterion for the scientists in the series, "Scientists of the Islamic Era". We state it briefly here. We have included only those scientists whose historicity is validated, those who have explicitly contributed through their writings and their works, and whose biography falls within the declared time period of focus in this series. Such criteria would not let us include mythical entries, and those without a trail of writings and works.

There is at least a three-fold purpose to this book series. One is to invite the world science community, in a manner of civilizational dialogue, to celebrate the science giants that Islamic Era has contributed to the growth of science and technology at its foundational level, as well as at the level of expanding its frontiers. Another is to remind the Muslims of their love for "science" which every man and woman must acquire; not for worldly prominence, but for a better humanity in a better world. One other objective is to join hands with the rest of humanity by satisfying the upwelling desire of the youth to know the truth about Muslim civilization and the excellence of their pursuit for wholistic knowledge: scientific, humanitarian, cultural, civilizational, and spiritual.

It is time for the world to move ahead of the historical biases, religious prejudices, cultural entanglements, and hegemonic aspirations. All people, together, constitute our humanity, and we hold this truth as self-evident that all humans are created with equal value. So, let us all join hands to bring science in the service of making every day a wonderful day in every neighborhood of our planet.

www.ingramcontent.com/pod-product-compliance
Lightning Source LLC
Chambersburg PA
CBHW051256120626
46547CB00015B/1971